DEFECT-FREE BUILDINGS

A CONSTRUCTION MANUAL FOR QUALITY CONTROL AND CONFLICT RESOLUTION

Books in the McGraw-Hill Construction Series

DEFECT-FREE BUILDINGS

A CONSTRUCTION MANUAL FOR QUALITY CONTROL AND CONFLICT RESOLUTION

Robert S. Mann
Construction Lawyer, Arbitrator, and Mediator

New York   Chicago   San Francisco   Lisbon   London   Madrid   Mexico City
Milan   New Delhi   San Juan   Seoul   Singapore   Sydney   Toronto

Copyright © 2007 by The McGraw-Hill Companies. All rights reserved. Printed in the United States of America. Except as permitted under the Copyright Act of 1976, no part of this publication may be reproduced or distributed in any form or by any means, or stored in a database or retrieval system, without the prior written permission of publisher, with the exception that the program listings may be entered, stored, and executed in a computer system, but they may not be reproduced for publication.

1 2 3 4 5 6 7 8 9 0   DOC/DOC   0 1 9 8 7 6

ISBN-13: 978-0-07-147959-2
ISBN-10:   0-07-147959-7

The sponsoring editor for this book was Cary Sullivan, the editorial supervisor was Jody McKenzie, the production supervisor was Jim Kussow, and the project manager was Vasundhara Sawhney. It was set in Times New Roman PSMT by International Typesetting and Composition. The art director for the cover was Anthony Landi.

Printed and bound by R R Donnelley-Crawfordsville

McGraw-Hill books are available at special quantity discounts to use as premiums and sales promotions, or for use in corporate training programs. For more information, please write to the Director of Special Sales, McGraw-Hill Professional, Two Penn Plaza, New York, NY 10121-2298. Or contact your local bookstore.

ABOUT THE AUTHOR

Robert S. Mann is one of California's leading construction lawyers, mediators, and arbitrators. He obtained the largest construction defect arbitration award in California history—$90 million—in a case involving the defective construction of the Shutters on the Beach Hotel in Santa Monica, California. During his 29 years of practice, he has represented some of the largest developers of multi-family housing in the United States in construction matters involving tens of millions of dollars. His private clients include the Marciano family, the owners of Guess? Inc., and many prominent members of the business and entertainment industry.

Mann is a member of the National Panel of Construction Neutral Arbitrators and Mediators of the American Arbitration Association, and he provides arbitration and mediation services through the AAA, ADR Services, Inc., in Century City, California and Forum Dispute Management in Los Angeles, CA. He is the principal of The Mann Law Firm, the practice of which is devoted to the litigation, arbitration, and mediation of construction matters. He has successfully tried more than 60 jury trials to verdict and has won many multi-million dollar verdicts and settlements. He lectures and writes extensively on construction issues and has successfully mediated more than 100 cases.

ABOUT THE CONTRIBUTING EDITORS

Robert McConihay is the principal of Construction Forensics, Inc., located in Newbury Park, California. He is a licensed general contractor with more than 25 years of experience in construction, forensic analysis, cost estimating, and reconstruction. He has served as an expert witness and consultant in hundreds of construction disputes.

Arron Latt is the principal of Project Management Collaborative, consultants to the construction industry in design, estimating, and construction management. A veteran in the construction business with more than 30 years of experience, Latt began his career as a facilities designer and planner, and he built a large firm providing design, architectural. and construction management services for in excess of 50 million square feet of commercial space.

Dean Vlahos is a principal of Widom Wein, Cohen, Terasaki & O'Leary, a leading architectural firm in Los Angeles. Vlahos, a licensed architect, limits his practice to forensic architecture. He has served as an expert witness and consultant in a variety of construction disputes. His expertise includes all aspects of building design and construction and the standard of care for architects and other design professionals.

Manny Shahraki is the principal of Los Angeles Soils, Inc. Shahraki is a licensed geotechnical engineer and provides soils engineering services and forensic services on geotechnical issues in construction matters. He has served as an expert witness and consultant on many significant soils-related construction disputes.

Contents

Acknowledgments

The first question that everyone asked after I told them that I wrote a book was: "Where did you find the time?" The truth is that I borrowed the time—from my family, my friends, and my other career as a construction mediator, arbitrator, and lawyer. So it's only fitting that I should first acknowledge and thank my wife, Mindy, and my daughter, Merissa, so very much for loaning me the time away from being a husband and father to complete this work. Without their encouragement and support, this project would have been impossible. I promise to repay the time, with interest.

I would like to express my deep appreciation to Cary Sullivan, my editor at McGraw-Hill, for her faith and confidence in this project. Without Cary Sullivan, this book would never have been published.

I am indebted to Jody McKenzie, Senior Project Editor at McGraw-Hill, for her patience and perseverance and to Vasundhara Sawhney, Project Manager, for her incredible attention to detail and her remarkable tolerance for my propensity to make "just a few more corrections."

To my contributing editors, for your wisdom, experience, and willingness to spend countless hours in making this work accurate and useful.

To my parents, for teaching me that great things are possible in life—you just need to apply the seat of the pants to the seat of the chair. To my brother, Charles (whose book on the economics of coal mining, also published by McGraw-Hill, is the definitive work in the field), because his ability to scientifically analyze just about everything has always amazed and inspired me. To my sister-in-law, Laura (an author of great mysteries) because she was the first person to read this book, and when she was done, she said, "It's great."

To professors Peter G. Arnovick, Ph.D., and Alfred V. Jacobs, Ph.D., with the deepest affection and utmost respect for inspiring me, teaching me, and giving me the confidence to go out in the world and pursue a life worth living.

To the experts, who for more than 25 years have taught me all that they know about construction, and to the clients who have entrusted me with representing them in memorable cases. To Lawrence Moon, for his assistance in helping describe this book in a way that made the publisher enthusiastic and interested and to Maria Valentine for so generously giving her information technology expertise.

How to Use this Book

This book was written to help construction professionals, owners and developers. Much of the information directed toward owners and developers is not technical enough to be of much use to construction professionals, and much of the information directed to construction professionals may be somewhat too technical for owners. For purposes of guidance, each section of this book is labeled *For Owners*, *For Contractors,* or *For Owners and Contractors.*

Although I have tried to provide the information to owners in clear, non-technical language, some information in the owners material may be of use to construction professionals. Similarly, even though the information for construction professionals assumes a certain level of skill and experience, owners can benefit from that information as well.

The sections for owners and contractors contain information that will be useful to both in the building process. Much of this information concerns communication and expectations. Some of the information also pertains to new technology in the construction industry, and it may be useful to both owners and contractors.

Much of the information with respect to the common types of construction defects may seem to contractors and other construction professionals to be simplistic, although it may seem overly technical to owners. To the contractors and construction professionals, I respectfully say that while the descriptions of the defects and the protocols for avoiding them may seem simplistic, it is nevertheless true that the professionals involved in construction litigation have seen every one of the defects that are described in this book countless times. This suggests that while it may be easy to avoid the defects, either contractors or other professionals are not focused on these potential problem areas or they are not paying attention to simple ways to avoid problems.

To the owners, I respectfully suggest that if the descriptions of the defects and protocols to avoid them seem too technical, use the protocols as a checklist for your contractor. As each aspect of the construction that is a possible source of defects begins to get underway, sit down with the contractor or subcontractor and discuss each of the items in the list and check them off, one-by-one, to insure that the best possible effort has been made to avoid the defects.

Introduction

If you are a contractor, architect, engineer, or developer, would you like to build a better product and reduce the chance of getting sued for construction defects or breach of contract? If you are a homeowner or developer, would you like to have a better built home, condominium, or commercial project and avoid lawsuits with your contractor? If the answer to these questions is yes, this book is for you.

After 27 years as a lawyer representing owners and contractors in construction disputes, helping parties resolve hundreds of such disputes in the capacity of mediator and deciding such disputes as an arbitrator, I realized that the same construction problems occur over and over again. I wrote this book because I believe that if owners and construction professionals had a better understanding of why construction problems arise, they could easily avoid those problems and the lawsuits that inevitably follow. By avoiding construction problems, contractors, subcontractors, and design professionals can make more money and build better projects. Owners and developers can have better buildings, and both sides in the construction process can avoid the cost and misery of litigation.

Many contract disputes between owners and contractors are the result of bad communication. Some disputes result from bad construction. This book will help you avoid disputes that result from both causes. If you should find yourself in a dispute, and if you have followed the advice in this book, you will be in a better legal position and it will cost you less in your lawsuit.

None of the recommendations in this book are intended to substitute for a contractor's years of experience in the building industry. Rather, these suggestions are intended to help contractors use their experience in a better way, to improve their business relationships, to enable them to make more money on their jobs, and to build better buildings.

Likewise, the recommendations will help owners and developers identify potential problem areas in advance of the construction process, manage the process more effectively and efficiently, and get better results without spending more money.

1 A Basic Primer on the Construction Process

FOR OWNERS

If you are like most people, you have neither built a new home, apartment building, or condominium nor remodeled an existing home. The construction process is likely a mystery to you. The selection of a contractor, negotiation of the contract with the contractor, the role of subcontractors, and issues of liens and insurance are confusing and difficult. Your lack of technical knowledge and information about the process of construction may make you nervous and you may think you are at the mercy of the contractor.

But in reality, in many respects, the construction process is not all that complicated. Understanding the basics of how homes and other wood frame structures are built, and understanding the fundamentals of the economics of construction and the relationship between the owner and the contractor, will allow you to move forward in the process with confidence.

Later in this book, in sections dealing with specific construction defects, you will read detailed discussions of various potential problem areas of construction. To put those into context, as an owner, you first need an overview of the whole process, with an explanation of some of the terms. So let's start with the basics and we'll get to the details later.

HOW WOOD FRAME STRUCTURES ARE BUILT

Virtually every wood framed home and simple wood frame structure in the United States is built the same way, using a process called *conventional light wood frame construction*. This type of building is called a *frame* or *wood frame* home. This type of construction is sometimes called *Type V* or *stick building*. Although many other types of

1

construction exist, some new and some old, most homes in the United States continue to be built with wood using techniques that have remained basically the same for hundreds of years. In fact, some familiar words and phrases are derived from centuries-old construction practices. For example, in medieval England, if you lived in a hut but could afford to keep a layer of wheat on the floor, you put a board across the entrance to your dwelling to keep the sheaves of wheat (the *thresh*) from falling out. The board was called a *threshold*, a term that we use today. If you weren't so fortunate, and your floor was bare dirt, you were "dirt poor," a condition in which you could find yourself today if your project ends up in a lawsuit.

Here's a basic description of wood frame construction: Every home starts with a foundation on stable soil or rock. Soil can be naturally stable, or it may be made stable by using various means to make it firm enough to bear the weight of the structure to be placed on the soil or to place structures through poor soil into stable rock under the soil. If the soil is not naturally stable and firm, those means include removing the existing soil and replacing it in layers while using machines to compact it firmly, or drilling large holes through unstable soil into the bedrock beneath the soil and filling those holes with reinforcing steel and concrete to create what are basically large concrete columns in the ground. The building is then supported on these columns, which are called *caissons, piles,* or *piers.*

Soil that is removed and properly re-compacted according to instructions given by a soils engineer or geologist is called *certified compacted fill.* It is called *certified* because the process of its compaction has been approved by the soils engineer and the engineer has tested the soil during the compaction process to verify that it meets certain minimum standards so that it will support the home or other structure to be built on the site.

Once it has been determined that the soil under the house is stable, a concrete foundation is built on the ground. The foundation is generally of one of two types: *slab on grade* or *raised.* Each type has its advantages and disadvantages (more about that later). A slab on grade foundation is made by digging a trench around the perimeter of the proposed structure. Wood forms are built inside the trench so that the trench has the shape of an upside-down *T.*

This upside-down *T* shaped trench is called a *footing.* Bars of round, textured steel (rebar) are laid down inside the forms to provide reinforcement for the concrete footing. Rebar comes in different thicknesses, each of which is numbered. The numbers correlate to an eighth of inch: thus, Number 3 rebar is 3/8 inch thick, and number 8 rebar is 1 inch thick.

In a slab on grade foundation, the dirt surrounded by the footings is leveled and prepared to accept more concrete. Concrete is then poured into the forms that have been built around the perimeter of the slab and, additionally, on top of the ground in the area surrounded by the footings. This type of foundation is called slab on grade because the concrete slab is poured directly onto the grade, meaning on top of the ground. Figure 1-1 shows a drawing of a slab on grade foundation and footing. The drawing shows how the foundation would look if you sliced through it (a section). The section shows the grade (the earth), 2 inches of sand, a layer of plastic (the vapor barrier), another 2 inches of sand, the concrete, and the rebar. The drawing also shows a *sill plate* (described in more detail soon) that is anchored to the foundation by an anchor bolt.

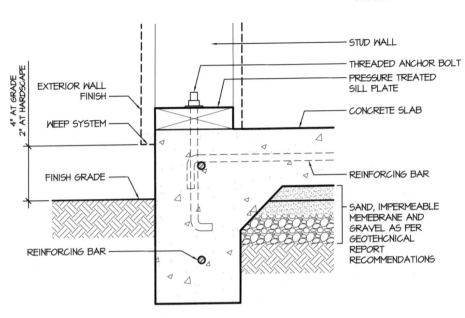

Figure 1-1

Slab on grade foundation and footing

STUD WALL

THREADED ANCHOR BOLT

PRESSURE TREATED SILL PLATE

CONCRETE SLAB

EXTERIOR WALL FINISH

4" AT GRADE
2" AT HARDSCAPE

WEEP SYSTEM

REINFORCING BAR

FINISH GRADE

SAND, IMPERMEABLE MEMEBRANE AND GRAVEL AS PER GEOTEHCNICAL REPORT RECOMMENDATIONS

REINFORCING BAR

Sometimes special types of slabs are used to provide extra strength for a foundation system. One of these systems is called a *post-tensioned* slab. The other is called a *structural* or *mat* slab. A post-tensioned slab is built in much the same way as a conventional slab on grade foundation. The difference is that steel cables are placed inside the area where the slab will be poured, and these cables are stretched after the concrete is poured and then allowed to return to their original length so that they pull the edges of the slab inward. A structural or mat slab is usually 6 to 12 inches thick and has substantially more reinforcing steel than a conventional slab.

A raised foundation is made by digging trenches around the perimeter as described above, and building wood forms into which concrete is poured. The difference is that the stem of the *T* is longer—about 24 inches. When the concrete is poured, a portion of the stem of the *T* remains above the ground. The house is constructed on these raised foundations (also called *stem walls* or *foundation walls*) by using pieces of wood called *floor joists* to span the distance across the raised foundations instead of pouring concrete on the ground in the area surrounded by the footings. The top of the footings is at least 18 inches above the ground. When the wood is placed across from footing to footing, this leaves a crawl space under the floor of the home. Figure 1-2 shows a section of a raised foundation footing. The drawing shows the sill plate, the floor joist, the rim joist (blocks of wood between the floor joists at the edge of the assembly), the plywood subfloor, and the sole plate upon which the walls will be constructed.

Some important facts must be considered with respect to building a raised foundation or a slab on grade foundation. Raised foundations offer room in the crawl space to install utility connections, such as electrical, plumbing, and heating and air conditioning. Slab on grade foundations require careful planning because arrangements must be made for openings for utility lines before the concrete slab is poured, and it is expensive to change these openings later. Generally speaking, slab on grade foundations are less expensive but offer the homeowner fewer alternatives and less flexibility for later installations.

Figure 1-2

Section of raised
foundation footing

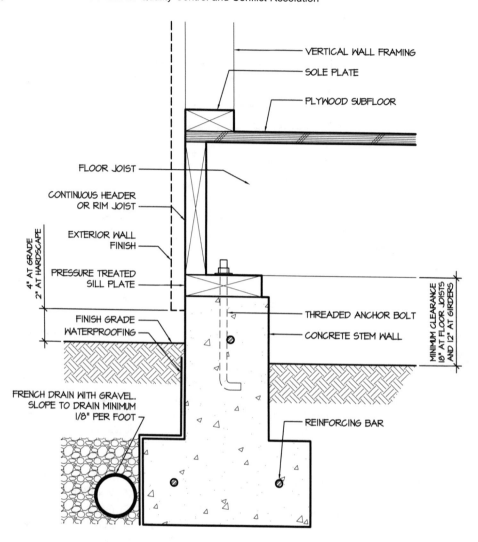

In areas where the soil under the slab is not strong enough to support the slab, as noted above, a series of caissons, or concrete columns, can be placed at intervals around the perimeter of the area where the foundation is to be built. Formwork (wood forms into which wet concrete is poured) is placed between the tops of the columns to create a horizontal concrete beam that is connected to the top of the caissons. These are called *grade beams* because the beams are placed directly on top of the grade. This type of foundation system is called *caisson and grade beam*.

Figure 1-3 shows a row of caissons that have been installed underground and are awaiting construction of the grade beam. The large vertical metal cage of rebar is typical of the reinforcing steel that is placed inside of each caisson prior to concrete being poured. The horizontal assembly of rebar will be attached to the top of the caissons and will reinforce the concrete to construct the grade beam.

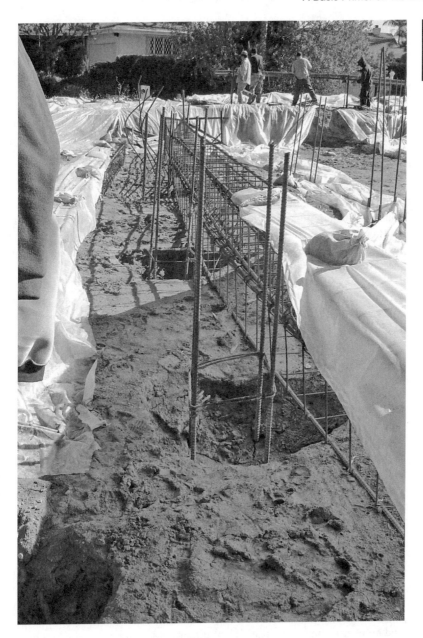

Figure 1-3

Caissons awaiting construction of the grade beam

During the construction of the foundations, and possibly even before they are constructed, depending on whether extensive soil preparation work is done, the foundation construction must be inspected by a building inspector. Building inspectors are employed by local governments (in small cities and towns, they are sometimes not government employees, but private individuals who contract with the city or town to provide inspection services). The building inspector's job is to see that the construction complies with the plans and specifications prepared by the architect and engineers and that it also complies with the building codes that apply to the location where the job is being built. Building inspectors are trained and experienced in construction, although it is not uncommon for them to miss many issues during the construction of a building. Most cities and towns also have a chief building official, who has the discretion to interpret the building code and is usually the final judge of what the municipality will or will not allow with regard to a particular construction technique.

Before the government's building department will allow construction to start, the owner must submit a set of documents to the building department to obtain a building permit. These documents generally include architectural plans (which show the basic design of the structure), structural drawings and calculations (prepared by a licensed structural engineer that provide details to make the building strong), soil reports from a geologist (to demonstrate that the soil under and around the structure is suitable), and energy calculations (to show that the proposed structure will comply with local and state energy requirements—so that it will not use too much energy). The owner, the contractor, the engineer, or the architect can submit the plans. The building department requires a plan check fee for reviewing the plans.

An individual in the building department, called a plan checker, reviews the plans, either over the counter (for simple structures) or more typically in his or her office over a period of weeks. The plan checker usually requires some corrections to the plans, and these corrections are communicated to the owner in the form of written comments or marks on the plans. The architect or engineer will either revise the plans, or, if they disagree with the building department, they may meet with the plan checker to discuss the proposed revisions. When the plan checker approves the plans, the building permit is ready to issue. Upon payment of the permit fee, the building department issues the building permit and work can begin.

To keep track of the inspection process, the building inspector issues a card to the contractor or building owner. This card is a form that has a list of items that are to be inspected. As the job progresses, the building inspector signs off on the various items. At the end of the job, the building inspector conducts a final inspection and gives a final approval for the work.

After the foundation is built, whether it is raised or slab on grade, the next step is to build the wood structure of the house. The wood structure portion of the house above foundations is called the *superstructure*. The wood superstructure of a home is made from *dimensional* lumber that is cut at a sawmill into standard sizes—such as a 2 × 4. The first number of the size indicates the thickness of the piece of wood, and the second number is the width. The size of dimensional lumber is called *nominal*. The nominal size is larger than the actual size, by a half-inch. Thus, a 2 × 4 is actually 1 $1/2$ inches thick and 3 $1/2$ inches wide. Other common examples of dimensional lumber are 2 × 6, 2 × 8, 2 × 10, 2 × 12, and 4 × 4. A 2 × 4 that is 8 feet long is called a *stud*. Dimensional lumber is used because wood frame construction is standardized so that all the pieces will fit together with a minimum of cutting and waste. Figure 1-4 shows some common sizes of dimensional lumber.

One important thing to understand about all wood products, including the dimensional lumber that is used to frame a house, is that wood is not stable. Wood expands and

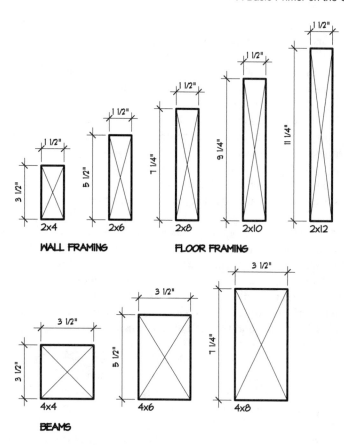

Figure 1-4

Dimensional lumber

contracts with changes in temperature and humidity. Wood will also bend and twist, both from the natural structure of the wood itself and from forces applied to it. Wood that is grown in managed forests is particularly susceptible to movement because the trees grow much more rapidly than old growth natural lumber. Wood that has a high moisture content from the lumber mill will also shrink, twist, or warp much more than wood that is drier. Wood can be dried at the lumber mill, in a large heated room called a *kiln* (kiln-dried lumber), or it can dry on-site after it is installed if the weather conditions permit.

Generally, lumber arrives at a construction site with about 18 to 19 percent moisture content. If the wood's moisture content is significantly greater, the moisture in the lumber could cause mold in the interior of the structure. Lumber that looks wet should be allowed to dry on-site before installation and certainly before it is covered on the outside or inside by drywall and exterior finishes. Optimally, the moisture content in framing lumber should be no greater than 14 percent after the framing is completed and before any finished surfaces are installed.

Over time, after the house is built, the moisture content will gradually fall to about 8 percent. Wood shrinks at a predictable rate, and generally in width, not length. The amount of shrinkage is surprising. A 2×12 will shrink in width approximately 5/16 inch as its moisture content falls from 19 percent to 8 percent.

The importance of this is that the movement of the wood, and its propensity to shrink and expand, can dramatically affect the finished result of a home. Walls can appear to be bowed. Cracks in wall coverings or paint can appear as the wood shrinks. The connection between foundation anchor bolts and the nuts that secure them can loosen as the wood shrinks. In recent years, a number of manufacturers have begun to create products designed to reduce or eliminate some of the natural movement of the wood. These products, which are generically called *engineered wood*, are made of small slivers or pieces of wood that are glued together under high pressures and temperatures. As a result, they are much more stable than natural wood. They are, however, more expensive than standard lumber.

To resist the force of wind and earthquakes (called *wind load* and *seismic forces*), the wood structure must be securely attached to the foundation and it must have a certain amount of stiffness. This is accomplished in several ways. First, pieces of wood called *sill plates* or *mud sills* are laid down on their sides on top of the slab if a slab on grade foundation is used or on top of the stem walls if a raised foundation system is used. Sill plates are usually 2×4 or 2×6 (in better construction) and are treated with chemicals to resist rotting and termite damage because they sit directly on concrete, which contains moisture. The chemicals are injected into the wood under pressure. For this reason, this type of wood is usually called *pressure-treated*. Pressure-treated wood or other moisture-resistant wood must be used anywhere it comes into contact with moisture. Pressure-treated wood is readily identifiable because a series of small grooves appear in the surface (it is usually greenish in color as well).

A note on pressure treated wood: In years past, wood was pressure treated with chromated copper arsenate (ACCA). Due to concerns about the toxicity of this material, ACCA has been replaced with treatments that use alkaline copper quat (AACQ), ammoniacal copper zinc arsenate (AACZA), and ammoniacal copper citrate (ACC), all of which are corrosive. These newer kinds of pressure-treated wood require use of particular types of metal anchor bolts and other metal devices. If the wrong type of metal is used, a chemical reaction can occur between the pressure-treated wood and the metal, and this can weaken or destroy the metal. Stainless steel, Z-MAX (a trade name for a Simpson Strong-Tie product that uses a thicker coating of zinc), USP Triple Zinc G-185 (another proprietary product), or post hot-dipped galvanized steel products can be used with newer pressure-treated materials. Copper, often used as a waterproofing material, can corrode if placed in direct contact with newer types of pressure-treated lumber and must be physically separated from the wood by a layer of some other material, typically plastic. An alternative is to use wood treated with SBX, sodium borate. Borate treated wood resists rot and insect damage, is non-toxic, and will not corrode metal fasteners. One caution: borate treated wood cannot be left exposed to weather for any considerable period of time because the borate will leach out of the wood.

New products promise to seal the gap between the top of the foundation wall and the bottom of the sill plate, to avoid the infiltration of air into the building structure. These products are made of rubber, vinyl, or foam, and they form a gasket between the concrete and sill plate.

The pressure-treated sill plates are bolted down to the foundation with *J*-shaped metal bolts called *anchor bolts*. The bottom portion of the bolt is smooth, and the top 2 inches has threads. The bottom portion of the anchor bolt is embedded into the wet concrete of the foundation at the time the concrete is poured. When the concrete cures (hardens), the anchor bolt will be firmly embedded in the concrete foundation system. Each anchor bolt is placed at a specified distance from another. An anchor bolt is usually about 8 inches long and about 3/4 inch in diameter. The top of the anchor bolt sticks up above the stem wall about 4 inches. A hole is drilled in the sill plate at the location of each anchor bolt. The sill plate is then lowered down over the anchor bolts, and a washer and nut is threaded onto each anchor bolt to tighten down the sill plate to the foundation. Figures 1-5 and 1-6 show a typical anchor bolt installation.

Figure 1-7 shows the installation of an anchor bolt in a remodel project. In this case, the anchor bolt has been installed before the concrete foundation has been poured.

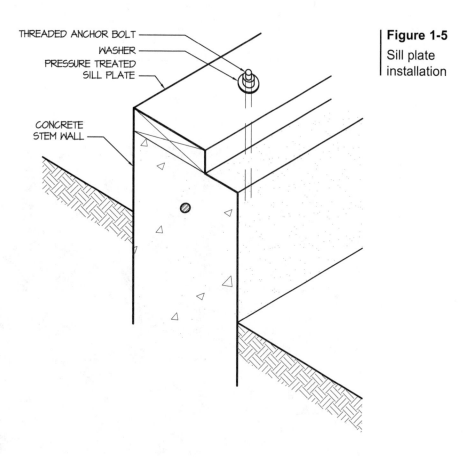

THREADED ANCHOR BOLT
WASHER
PRESSURE TREATED
SILL PLATE

CONCRETE
STEM WALL

Figure 1-5

Sill plate installation

Figure 1-6

Sill plate bolted
to stem wall

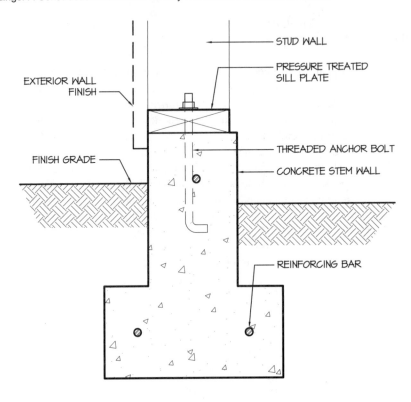

STUD WALL

PRESSURE TREATED
SILL PLATE

EXTERIOR WALL
FINISH

THREADED ANCHOR BOLT

FINISH GRADE

CONCRETE STEM WALL

REINFORCING BAR

Figure 1-7

Anchor bolt
installed before
foundation poured

To support the floor, pieces of wood are installed across the foundations, on top of the sill plates. These floor joists are set on edge on top of the sill plates. The ends of the floor joists are set back from the outside edge of the sill plate by 1 $\frac{1}{2}$ inches. A piece of 2-inch-thick dimensional lumber is then placed on edge on top of the outside edge of the sill plate to cover the exposed ends of the floor joists. This piece of lumber (the size of which will vary according to the height of the floor joists) is called a *rim joist*. The floor joists and rim joist together are called the *floor joist assembly*.

A plywood *subfloor* is placed on top of the floor joist assembly. Another plate, called the *sole plate*, is nailed on top of the subfloor at the edge of the foundation. The wood walls are then attached to the sole plate. Figure 1-8 shows the foundation, the floor joists, the rim joist, a portion of plywood subfloor the sole plate, and a portion of the vertical wood framins.

The walls are also made of dimensional lumber. In a conventional home, the walls are made mostly of 2 × 4 studs. In better construction, the walls are made of 2 × 6 studs so that thicker insulation can be placed in the walls. The walls are called *stud walls*.

The stud walls are attached to the sill plate on top of the floor joist assembly using nails and specially manufactured metal devices called *hold-downs*. These metal devices are used to fasten the stud walls securely to the sill plates so that the house will resist seismic and wind forces. These metal devices are anchored in the concrete stem walls and then bolted to the studs and the sill plates to make a strong connection. The leading

Figure 1-8
Floor joist and framing assemblies

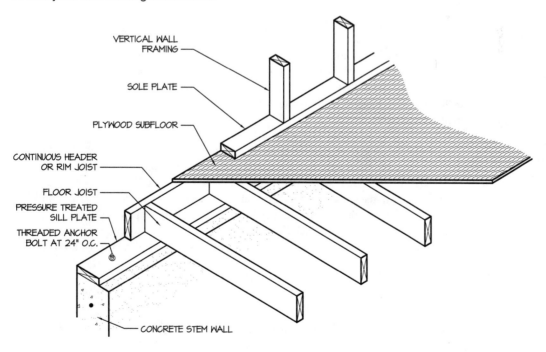

VERTICAL WALL
FRAMING

SOLE PLATE

PLYWOOD SUBFLOOR

CONTINUOUS HEADER
OR RIM JOIST

FLOOR JOIST

PRESSURE TREATED
SILL PLATE

THREADED ANCHOR
BOLT AT 24" O.C.

CONCRETE STEM WALL

Figure 1-9

A typical hold down

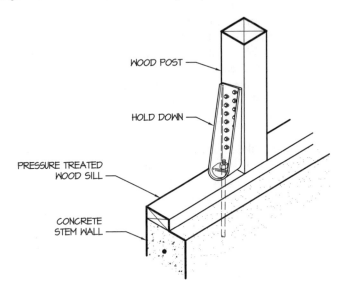

WOOD POST

HOLD DOWN

PRESSURE TREATED
WOOD SILL

CONCRETE
STEM WALL

manufacturer of these metal devices in the United States is the Simpson Strong Tie Company, which makes a wide variety of structural metal devices, including straps, clips, hold-downs, and other specialty materials. Figure 1-9 shows a typical hold down that is usually installed at the ends of a wall.

At the top of the wall, two pieces of wood are laid on their side, one on top of the other. These are called *top plates* or a *double top plate*. If the stud wall is made from 2 × 4s, the top plates will also be 2 × 4s. If the stud walls are 2 × 6s, the top plates will be 2 × 6s. To make the ceiling, additional pieces of wood, usually 2 × 10s, are laid on their edges on top of the top plates. These are called *ceiling joists*.

Figure 1-10 shows an entire exterior framed wall assembly. Starting from the bottom of the photograph, one can see the top of the concrete foundation wall, the sill plate, the vertical studs, a window opening (over which is a large piece of framing material called a "header") engineered lumber at the top of the wall ceiling joists and fireblocking (see explanation below).

In a raised foundation system, floor joists, usually 2 × 10s or 2 × 12s (depending on the distance that the piece will have to extend without being supported underneath, called the *span*) will span the distance across the raised foundations. Floor joists are placed on their edges, with the ends sitting on top of the sill plates. They are spaced 12 to 14 inches apart, depending on the length that they must span. Additional pieces of wood the same size as the floor joists are placed on top of the sill plates, between each of the floor joists, so that the floor joists are braced and remain straight up and down. These pieces of wood are called *blocks*.

After the floor joists are installed, sheets of plywood are nailed (and in good construction, glued as well) over the floor joists to make a subfloor or deck. In a slab on grade foundation, floor joists are unnecessary at the first floor because the concrete slab on the ground forms the first floor (although a plywood subfloor may be installed to

Figure 1-10
The exterior framed wall assembly

provide a base for hardwood or other finished flooring surfaces). Two kinds of plywood can be used to make the subfloor or deck. Standard plywood, or tongue-and-groove plywood. Standard plywood has square edges, and tongue-and-groove plywood, which is more expensive, has a slot, or groove, cut into one edge, and a small ridge, a tongue, on the other edge. As each sheet of plywood is installed, the tongue from one sheet is fitted into the groove of the next sheet. The tongue-and-groove configuration makes the subfloor much stronger and more resistant to movement. Regardless of what kind of plywood is used, the plywood sheets are glued and either nailed or screwed to the floor joists so that the entire assembly is firmly connected together to reduce movement and noise (from squeaks).

As noted, sometimes, depending on what material will be used for the finished floor surface, a plywood subfloor is installed over the concrete slab. When this occurs, a plastic sheet, called a *vapor barrier*, is installed between the concrete and the subfloor to prevent moisture and water vapor from being transferred from the concrete to the wood subfloor and the finished floor surface above it.

In recent years, manufacturers have developed new products that are used in place of conventional floor joists. These products, a form of engineered lumber, are called by different trade names, but most are a form of truss joist and they all work in much the same manner. They are shaped like an *I* and are made of various materials. Usually the vertical portion, called the *web*, is made of pieces of chipped wood that are glued together under high temperature and pressure. Sometimes the web can be made of metal tubes. The horizontal top and bottom pieces of wood, called the *flange*, are made of small layers of wood, glued together in plies, under heat and pressure (much like plywood). The usefulness of all the various types of these new products is the same: they are light, not prone to shrinkage, stronger than conventional dimensional lumber, able to span longer distances than conventional lumber, and much more stable than conventional floor joists, so that the floor above them will move less and will be flatter and quieter. They have one additional benefit: the web, if made of wood, has wood plugs installed, called *knockouts*. When the plumber or electrician needs to make a hole in the flange to run conduit or pipe, he or she merely hits the plug with a hammer and the plug pops out of the web. This eliminates the need to drill holes in the webs, which saves considerable time and effort.

Figure 1-11 shows the use of engineered truss joists in the ceiling of an elaborate custom home. The photograph also depicts the installation of a complex framing structure to form coves for the ceiling of the room, directly below the truss joists.

If the house has a second story, another set of stud walls is built on top of the top plates and another set of ceiling joists is installed above the second floor walls. In either case, a wooden roof structure is built using various means. Some wooden roof structures are made on the site by the framing contractor, and some are made in factories and are brought to the site and installed.

The basic assembly of sill plates, studs, top plates, floor joists, and ceiling joists is not strong enough to withstand seismic (earthquake) or wind forces. To make the house strong enough to withstand these forces, sheets of plywood are nailed to the outside and sometimes to the inside walls. These sheets of plywood are required to be a certain thickness. In addition, a specific thickness and length of nail is required to be used, and the nails are required to be spaced a certain distance apart along the edges of each sheet of plywood and elsewhere on each sheet of plywood. These specially installed sheets of

Figure 1-11

Truss joists in a ceiling

Figure 1-12
Shear wall
assembly
for two-story
exterior wall

plywood are called *shear panels*. The name derives from the fact that lateral forces applied to a structure tend to move or "shear" the structure from the foundation. Shear panels can also be made from manufactured lumber call Oriented Strand Board, or OSB, which is made by gluing chips of wood into panels under heat and pressure.

Figure 1-12 shows an exterior two-story wall. The bottom, or first story, has been covered with shear walls. The shear wall assembly has started to reach into the second story.

Another important element of the basic framing is *firestopping*. Recall that the vertical walls, the stud walls, are built using 2×4 or 2×6 studs, spaced every 16 inches apart. On the outside, the stud walls will be covered with shear panels and the final exterior finish (such as stucco). On the inside, the stud walls are usually covered with drywall. If a fire were to start and travel up the inside of the walls in the cavity formed by the exterior and interior, it could quickly reach the top plates and roof of a single-story home or the top plate and second story of a two-story home. If the fire is not interrupted, it can travel upward to the attic space and set the roof on fire. To slow the spread of the fire, pieces of wood called *fire blocks* are installed horizontally between each of the wall studs to block the cavity. The concept is that a fire traveling upward through the cavity between the studs will take at least one hour to burn through the fire stop. By that time, it is hoped that someone will have recognized that a fire has started to put it out before additional damage is done. These types of walls are called *one hour walls*. Figure 1-13 shows fire blocking within a frame wall.

Figure 1-13

Fire blocking
slows a fire from
traveling upward

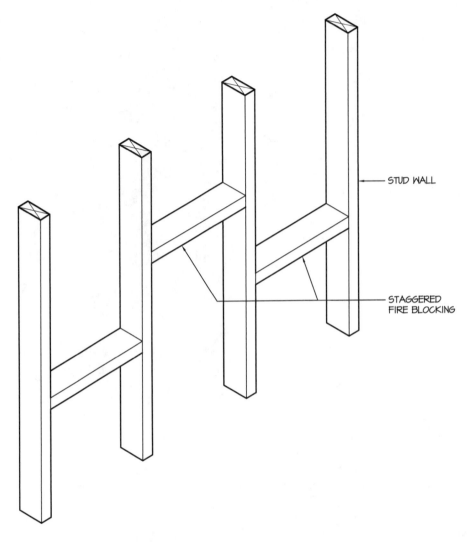

STUD WALL

STAGGERED
FIRE BLOCKING

The basic framing of a home described so far is called *rough* framing, and it is completed when all of the floors, walls, ceiling joists, and roof framing assembly (the roof raming and sheathing but not the actual roofing material, such as shingles or tiles) have been installed. At that point, the building inspector will be called to inspect the rough framing.

The inspector will review the plans to determine whether all the various details required by the engineer have been met. He or she will look for the hold-downs and other important structural elements, including the installation of the shear panels. When the inspector is satisfied that the rough framing is complete, he or she will approve the rough framing in writing by making a notation on an inspection card and the job can progress to the next step.

During the construction of the walls, openings are left for windows and doors. A horizontal piece of wood, called a *header*, is installed over every such opening to support the weight of the load above the opening without letting that weight rest on top of the door or window. The horizontal piece of wood at the bottom of the opening is called a *sill*. The horizontal piece at the bottom of a door opening, particularly when finished, is called a *threshold*. Figure 1-14 shows a typical window opening.

Windows and doors can be made from various materials. The most common are all wood windows, windows that are wood on the inside but covered with metal or vinyl on the outside, windows that are constructed from metal, and windows that are constructed from solid vinyl. Windows can also be made of solid aluminum and steel. Various kinds of glass can be used to provide for greater insulation, resistance to ultraviolet rays of the sun, and even to resist the build-up of dirt on the outside of the window.

Similarly, doors can be built of wood, wood and vinyl, wood and metal, or all metal. Doors and windows require a seal to prevent wind, cold, and heat from getting into the home. This is called *weatherstripping*, and it can be constructed of many different types of materials, including vinyl and metal. It is important to distinguish between weatherstripping and waterproofing. Weatherstripping is primarily intended to stop air

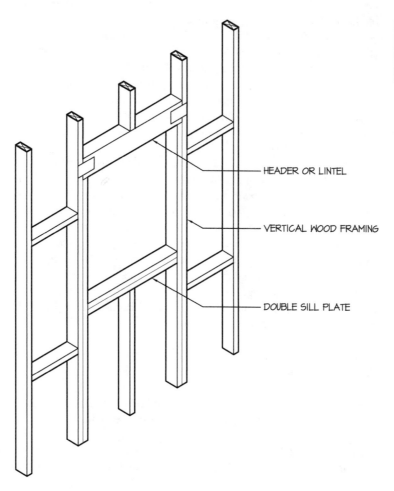

Figure 1-14

Window opening with header and sill

HEADER OR LINTEL

VERTICAL WOOD FRAMING

DOUBLE SILL PLATE

movement around doors and windows to prevent drafts. It has a secondary benefit of resisting moisture, such as wind-driven rain. Waterproofing is the primary means of defense against water intrusion and is intended to keep water from rain and snow from making its way from the outside of the home into the wood framing or interior.

The windows and doors are installed into the openings and waterproofing is attached to them so that water does not run into the openings and into the inside of the home or into the stud walls. This is a critically important part of the construction process. It is estimated that more than 75 percent of all construction defect cases start with water intrusion, and most water intrusion comes from faulty installation of waterproofing around windows and doors. Later, in Chapter 5, you'll read about waterproofing around windows and doors.

Now that the basic structure is built, the utilities have to be provided to the home and the home must be finished. The basic utilities are water, sewer, and electricity.

To wire a home for electricity, a wire is run from the electrical utility company source in the public roadway to a metal box (an *electrical panel*) mounted on the outside of the house. The panel contains an electrical meter that records the amount of electricity used by the homeowner. From the electrical meter, wires are run to a metal bar called a *bus bar*, which is used to distribute the electricity through the house by using circuits and circuit breakers. A circuit is simply a separate line of electrical wiring that will provide power to a specified number of electrical outlets, lighting fixtures, or appliances. Each circuit starts with power being supplied to a circuit breaker. The circuit breaker is attached to the bus bar. The circuit breaker functions as a fuse to shut off the flow of electricity if the circuit has too much electrical current running through it. Older houses may have fuses instead of circuit breakers: they both serve the same function.

The wires for each circuit are run into the various rooms of the house by cutting holes in the studs, sill plates, and joists and running a metal tube called *conduit* through the holes. Conduit can be flexible or rigid, and it is made of various materials, but usually aluminum inside a home and plastic outside a home. In some houses, conduit is not used. Instead, wires that are coated with heavy plastic insulation are used. The common trade name for this product is Romex. When conduit is used, the electrical wires are pulled through the conduit and attached to receptacles, outlets, and switches. The advantage of conduit is that the wires can be changed without opening up the walls by pulling new wires through the existing conduit. Romex, on the other hand, cannot be changed without tearing up walls and ceilings. Romex is usually used to save costs and is typically found in less expensive houses.

Figure 1-15 shows several conduit runs leading to a *J box*, or junction box, in which electrical receptacles will be installed.

The wires for switches and outlets end in metal or plastic boxes that are secured to the wall studs. These boxes will eventually contain the switches and outlets themselves.

The electrical installation has two main parts: the rough electrical and the finish electrical. The rough electrical consists of installing the electrical panel and connecting the house to the

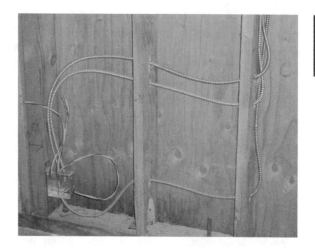

Figure 1-15
Conduit runs leading to a junction box

electrical utility lines, installing the circuit breakers, running the conduit, and pulling the wires through the conduit into the electrical boxes or running Romex to the boxes. The finish electrical consists of installing the switches, outlets, and light fixtures. When the rough electrical is completed, it must be inspected and approved by the building inspector. The finish electrical is also inspected and approved in writing on the inspection card.

Water comes to the home from a large pipe called a *water main*, usually under the nearest street, which brings the water to the home. A main valve connects the city water main to a water meter that records the amount of water used by the homeowner. The meter, in turn, is connected to a main line that runs to the house. Usually, a pressure regulator is installed to reduce the water pressure to between 60 to 90 pounds per square inch (psi) of pressure so that the plumbing pipes and fixtures inside the home do not burst from too much pressure. The water lines inside a home are generally made from either galvanized steel or copper. More recently, in some areas, PVC plastic pipe is used. In rare instances, plumbing pipe is made from brass. Galvanized steel will rust and corrode over time. Copper, if properly installed, has a long useful life, at least decades. Plastic pipe has a long useful life, but in the 1980s, lawsuits were filed alleging that a large quantity of PVC plastic pipe was manufactured improperly and failed in use. This, along with resistance from plumbing contractors, has limited use of PVC pipe in residential construction. In addition, unless specially made to resist ultraviolet rays, in which case it is labeled "UVB," PVC pipe will become brittle and crack when exposed to the sun and must therefore be shielded from direct sunlight.

In recent years, a proprietary flexible plastic pipe product called Pex has begun to emerge as an alternative to copper or steel piping. Pex has been used in Europe for many years and is starting to gain acceptance in North America. In a system that uses Pex, a large metal pipe with a number of connections, called a *manifold*, is installed. Pieces of the Pex tubing are connected to this manifold and then directly run to each of the appliances, such as sinks, showers, and tubs. The advantage of Pex is that it can be installed quickly because it does not require time-consuming soldering to make joints, as is required for copper piping. The material itself and the fittings, however, make a Pex installation about the same price as a copper pipe installation.

Like the electrical wiring, the plumbing pipes are run through holes in the studs and joists. Every plumbing fixture must have a vent line and a drain line. Drain and vent lines can be made from ABS, a heavy black plastic; PVC, a lighter plastic; or cast iron. Vent lines are usually ABS or PVC. The use of cast iron drain lines will reduce noise considerably and is recommended in better construction. In some instances, cast iron will be required by local building codes. Vent lines are used to vent sewer gas to the outside atmosphere and to eliminate the vacuum that is formed by water running down the drain line. The vent lines are installed inside the walls and penetrate the roof. The drain lines drain water and sewage to either a city sewer system or a septic tank and leach field. A septic tank and leach field is basically a private wastewater treatment facility that is buried in the ground near the home. A septic tank and leach field is used in areas that are not served by a municipal waste water system (a sewer system).

Like the electrical system, plumbing has both a rough and a finish component. The rough plumbing is the running of the pipes to the various outlets, such as showers, toilets, sinks, hose bibbs, and hot water heaters. The finish plumbing is the installation of faucets, toilets, sinks, shower heads, and valves. The rough plumbing must be inspected by the building inspector. When the building inspector makes the inspection of the rough plumbing, all of the plumbing lines are capped and the lines are pressurized, meaning that water from the main water line is allowed to flow into the pipes under pressure. The inspector checks to see that pressure remains, which means that no leaks are present.

The vent lines also require inspection. To inspect the vent lines, all of the drain lines are plugged and the vent lines are filled with water. The water is left in the vent lines for at least 48 hours and then inspected by the building inspector.

At the same time, or usually at about the same time, the inspector will want to inspect any *shower pans*, a waterproof assembly for a shower floor. A basic shower pan is installed by a subcontractor called a *hot mopper*. It is usually made as follows: a drain hole is cut into the wood floor in the center of what will become the shower. A metal drain assembly is installed in the hole. Pieces of wood are installed on the floor to make a curb around the shower. Mortar (a mix of sand and cement) is installed inside the curb to the drain so that the floor of the shower will slope toward the drain. Blocks of wood at least 12 inches high are installed between the studs on the sides of the shower where there are stud walls to form a solid wall surface.

The hot mopper then lays a sheet of asphalt-impregnated felt paper (similar to the material used for roofing) into the floor of the shower and over the curb. The hot mopper then coats the paper with hot melted tar and mops the tar up the sides of the walls about 9 to 12 inches. Additional layers of paper are installed with hot melted tar between each layer. A final coat of melted tar is applied over the floor, curb, and wall. When the tar cools, it forms a monolithic (one piece) waterproof coating. Better practice for the waterproofing of showers also involves installing a waterproof membrane at least 6 feet on the vertical walls of the shower enclosure (above the shower head) and lapping this membrane over the hot-mopped waterproof coating. This vertical waterproof membrane will prevent water that penetrates grout joints from damaging the structure.

To inspect the shower pans, the drain is plugged, the shower floor is filled with water to the top of the curb, and the water is left to stand for 48 hours. If the water level has not changed, the inspector will know that the shower pan does not leak.

There are several alternatives to hot tar shower pans, including prefabricated plastic or fiberglass plans, or installations that use various liquid waterproofing materials (other than tar) or plastic sheets that are glued together.

When the finish plumbing is completed, it must be inspected and approved by a notation on the inspection card.

A heating, ventilation, and air conditioning system (hvac system) is installed. Conventional (non-solar) heating systems use two methods to heat the living space: (1) radiant heat, using electrical wires or hot water; or (2) hot air, using electricity or gas to heat the air. Radiant heat can come from electrical wires or hot water pipes in the floors, in radiators, ceilings, or in baseboards. Hot air systems use a furnace or a hot water heater to heat the air, which is then distributed to the rooms through tubes called *ducts*.

A heating system that uses hot water to heat the air is called a *hydronic* system. In this kind of system, hot water is heated in a hot water heater or boiler. The hot water is piped to a heat exchanger (similar to an automobile radiator, but working in reverse). Air is blown over the heat exchanger, which has been warmed by the hot water. As the air passes through the heat exchanger, heat is transferred from the metal heat exchanger to the air. After the air is heated, it is forced through the ducts into the rooms in the house. The advantage to a hydronic system is twofold: it is efficient (because the newer technology hot water heaters are more efficient than even efficient conventional furnaces) and the air retains more of its moisture so that in winter the warm air is much less dry.

Solar heating systems can be used to heat water for radiant heating systems that use pipes in the floors or passive solar systems heat the actual building components (i.e., walls, floor slabs), which then radiate the heat throughout the structure. Solar heating systems are also commonly used to heat water for swimming pools and for domestic hot water use.

Air conditioning systems are either centralized or room type. Room type systems use an air conditioner mounted in the wall. Central systems use a component system to chill the air, which is then distributed to the rooms through ducts. When a central heating and air conditioning system is used, the same ducts are generally used to distribute both hot and cold air.

The hvac system will require approval at the time the plans are submitted so that the building department can confirm that the system meets energy-use guidelines imposed by the local municipality or the state. The system will also have to pass inspection by the building inspector. Once the rough framing, rough electrical, and hvac systems have been inspected and approved, the contractor will have approval to install insulation.

All homes require insulation to maintain as uniform a temperature as possible throughout the year. (Under California law, for example, a heating system is required to maintain a constant temperature of 70 degrees 3 feet from the floor in any living space and if it cannot do so, it is deemed to be defective construction. California Civil Code section 896(E)(4).) Various forms of wall insulation can be used. The most common is fiberglass insulation, which comes in the form of large "batts" that are designed to fit between the vertical wall studs and between ceiling and floor joists. A batt of insulation is a piece of fiberglass that is cut into a certain size and thickness and sometimes covered with plastic or paper on one or more sides. Other forms of insulation include liquid foam, which expands and hardens after it is sprayed into the cavities between the studs and joists; loose cellulose insulation, which is blown into the cavities; and batts of insulation that are made from materials other than fiberglass, such as recycled cotton fabric. Rigid foam insulation may also be used in walls and in attic spaces.

Figures 1-16 and 1-17 show the installation of fiberglass batt insulation in the wall and ceilings of a residence.

Figure 1-16

Fiberglass batt insulation in walls and ceiling

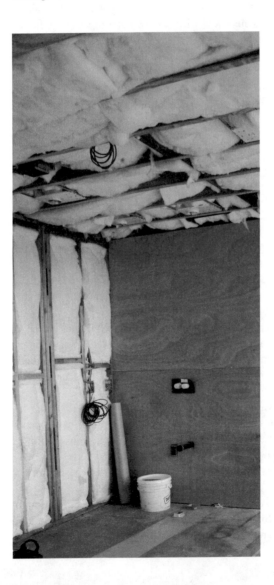

Figure 1-17
Fiberglass batt insulation in the wall

Insulation is also used in basement walls and at the roof level of homes, particularly in hot climates. Roof insulation applied to the surface of the roof (directly under the roofing material) is almost always made from foam—either rigid foam, which is supplied in large panels, or foam that is sprayed onto the roof using special equipment. Basement insulation is usually also foam, in the form of panels that are attached to the basement walls underground.

The building code spells out the minimum insulation required. The building inspector will inspect and approve the insulation before the next step in the process.

Before the walls can be covered, the interior door jambs must be installed. When the house was framed, openings were left for the interior doors, but these openings are not the final size or finish and are called rough openings. To finish these openings and install the interior and exterior doors, two pieces of wood must be installed vertically inside the rough opening and one piece of wood installed horizontally at the top of the rough opening. This assembly is called the *door jamb*. The interior and exterior doors are then attached with hinges to the door jambs.

After the door jambs are installed, the interior walls are covered with gypsum wallboard (commonly called *drywall*) that is nailed or screwed to the inside of the wood frame to make the finished wall surfaces of the rooms. Sometimes a different system, lath and plaster, is used. *Lath* is metal, wood, or button board (a type of plaster board that has holes through which the plaster intrudes into the wall cavity) that is attached to the framing. *Plaster* is then troweled onto the lath and finished to a smooth and flat surface. True lath and plaster is quite rare in new construction, particularly in the Western United States.

The installation of the drywall must be inspected by the building inspector, who will look to see that the drywall has been screwed or nailed properly (when installing drywall, the screws and nails must be placed a certain distance apart and in certain locations and the screws must not be driven too far beyond the surface of the drywall) and to determine whether the drywall has been finished properly.

In areas with high moisture content, such as bathrooms and laundry rooms, standard drywall is not used. In past years, a material called *greenboard* was used in these areas. Greenboard is substantially similar to drywall, but the paper coating is made to be water resistant. I do not recommend the use of greenboard, however, because it has been my experience that greenboard will not, in fact, resist moisture and will gradually deteriorate over time. In an effort to resolve issues with greenboard, several of the large gypsum wallboard manufacturers have begun to make different products. Georgia Pacific, a manufacturer of drywall products, makes several types of water-resistant drywall under various trade names. These products are basically a form of gypsum coated with fiberglass instead of paper. The cost of using fiberglass-wrapped gypsum wallboard is not significantly more than using standard greenboard, and these products will perform in a much improved manner.

On the outside of the home, a variety of coverings can be used, such as wood siding, brick, or cementitious exterior plaster (commonly called *stucco*). All of the exterior coverings share one thing in common: the home must be waterproofed before they are installed so that water will not come into contact with either the wood frame of the home or enter into the interior of the home.

If the exterior walls are to be covered with stucco, an additional inspection is required—a lath inspection. Lath is used to bond exterior plaster to the house mechanically. The lath is made of a wire grid. The installation is done as follows: The window and door openings must be waterproofed. After the windows are installed and waterproofed, the stucco subcontractor wraps the outside of the house with a special type of black paper called builder paper, which is stapled onto the outside of the framing. The builder paper allows any water that penetrates the stucco (because stucco is not waterproof) to run down the side of the building and drain out to the ground before it can come into contact with the wood framing of the house. The builder paper is therefore lapped, or shingled, like a roof, so that water will run down and not become trapped between the layers of paper.

Figure 1-18 shows factory made panels that combine builder paper and wire lath that have been installed on the exterior of a residence. In the top photograph, one can see the layering, or shingling effect of the panels.

After the builder paper is installed, various types of wire mesh are nailed or stapled to the building (unless panels of builder paper and metal lath are used). This wire mesh is the *lath*. At the bottom of the walls, not less than 6 inches above grade, a metal device called a *weep screed* is installed. This serves two functions: (1) it serves as a screed, or straight edge, so that the stucco is straight and flat; and (2) it permits water that penetrates into the stucco during rainstorms to "weep" out of a series of holes so that the water does not remain in the stucco. Figure 1-19 shows a weep screed.

When the first coat of wet plaster is applied, some of the plaster is forced under and around the wire lath, and when it dries, it makes the plaster adhere firmly to the wire. Because stucco is quite heavy, it is critically important that the lath be attached properly and that the first coat of wet plaster be attached properly to the lath. Properly applied stucco is *keyed* into the lath. If the stucco is not keyed into the lath, or if the lath is not

Figure 1-18

Building paper and wire lath installed on the exterior

Figure 1-18
(*Continued*)

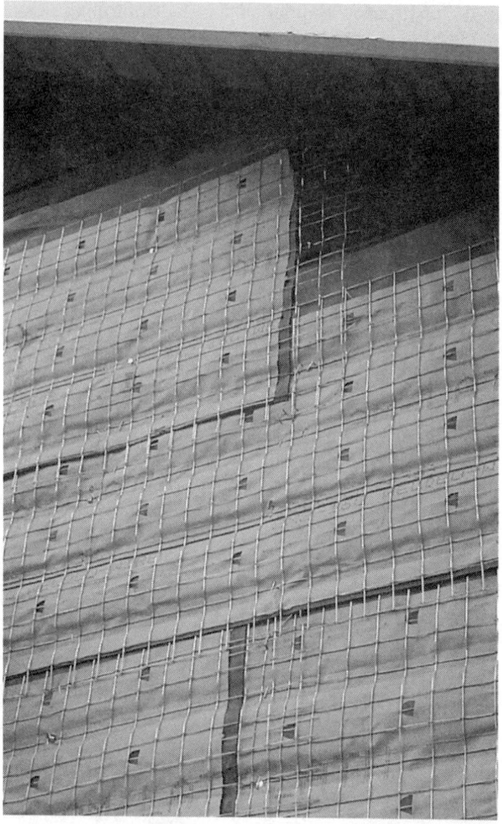

Figure 1-18
(*Continued*)

Figure 1-19
Weep screed at the bottom of a wall

Figure 1-20

Improperly embedded plaster in lath

secured to the building, the stucco can crack or peel away from the building. Figure 1-20 shows a piece of stucco that has not been properly embedded into the lath.

Two important issues are involved in attaching the lath properly. The lath must be nailed or stapled to the studs, not the plywood covering on the outside of the house. The staples or nails must be embedded into the studs at the depth required by the building code. This means that the stucco subcontractor must use nails or staples that are long enough to grip tightly into the studs to resist the weight of the stucco.

When the first coat of wet plaster is applied to the lath, it must be forced under and around the wire lath so that it is firmly connected. If not, the stucco can peel away from the building or crack prematurely. When the building inspector inspects the lath work, he or she will be looking for proper lapping of the builder paper and proper attachment of the lath.

Most stucco systems are three coat systems. The first coat, or "scratch" coat, is applied and then scratched with a special tool that resembles a small rake to create deep grooves. Figure 1-21 shows the "scratch coat" on a stucco wall.

The scratch coat is allowed to dry for at least two days, but preferably for at least one week. After the scratch coat dries, a "brown" coat is applied. The brown coat is thinner, and it is finished with a wood trowel to give it a flat but rough surface. The brown coat is also allowed to dry, preferably for at least one week. The final coat of stucco is then

Figure 1-21
The scratch coat

applied to the brown coat. If the final coat has color mixed into the material, no painting is required. Otherwise, paint is applied over the final coat.

Figure 1-22 shows a section of a stucco installation (seen from the end of a stuccoed wall). Visible in the photograph are two layers of builder paper, metal lath, and three coats of stucco.

A roof covering is installed over the roof system. This covering can consist of a variety of products, such as asphalt shingles, clay roof tiles, stone roof tiles, metal roof systems, and "built-up" roof systems.

Built up roof systems are required when the slope of the roof is nearly flat. A common built-up roof system uses three to four layers of material, with hot melted tar between each of the layers. The first layer of material is a fiberglass material called a *base sheet*, which is nailed to the plywood roof sheathing. A layer of hot melted tar is mopped over the base sheet, and a layer of asphalt felt roofing material called an *interply* is laid on top of the melted tar while it is still hot and liquid. Another layer of melted tar is applied, with another interply placed on top of the melted tar. Two or three interplies can be placed on the roof. Finally, a final layer of a different material called a *cap sheet* is laid onto the last layer of melted tar. The cap sheet comes from the factory with a surface that is covered with granules of sand in different colors and it is strong enough to walk on (although walking repeatedly on a cap sheet will damage the surface and will cause premature wear: to avoid this problem, pads of a resistant material are placed on the roof in areas where people will walk on the roof surface). There are other roofing systems for low slope roofs, and many varieties of roofing materials and systems can be used for roofs that have a steep slope.

Figure 1-22

A section of stucco showing two layers of builder paper, metal lath and three coats of stucco

Vent pipes of black ABS plastic, sheet metal vents, and chimneys, whether made of brick, metal, or wood covered with stucco, all penetrate the roof. Anything that penetrates the roof must be waterproofed so that water does not flow through the space between the hole in the roof and the vent and then into the house. Various methods of waterproofing are used to seal these openings, but the important issue is that they must be installed with considerable care and attention to detail to avoid serious water intrusion problems.

After these basic construction activities are completed, the shell of the home is done. What is left are finishes, such as hardwood floors, carpet or vinyl, paint, paneling, wallpaper, fixtures, appliances, and cabinets. Floors are installed inside the house. Interior and exterior doors are installed. The inside and outside are painted.

When all of this work has been done, the building inspector is called for a final inspection. If everything passes inspection, the inspector approves the installation and issues a certificate-of-occupancy (C-of-O). This document allows the homeowner to move into the house and start enjoying the new home.

At the simplest level, this provided a good overview of how a house gets built. But between the first shovelful of dirt and the last coat of paint are hundreds of things that can, and often do, go wrong. To understand what can go wrong, and how to stop it before it happens, a much more in-depth discussion of the building process is required.

WHO IS INVOLVED IN THE PROCESS

It takes many people with many specialized skills to design, engineer, and build a home. Before a home can be built or remodeled, a design is created either by a designer, the owner, an architect, or any combination thereof. A designer and an architect are different. An architect has certain professional training and has a license from the state in which he or she practices. A designer may have many of the same skills and experience as an architect, but he or she does not have a license. Under the laws of many states, only architects can design certain kinds of buildings, such as hospitals and schools. Both a designer and an architect share at least one set of skills, and that is the ability to visualize an overall design concept for a new home or a remodeling project. The design concept includes the size and shape of the inside and outside and a sense of the materials to be used.

The architect or designer must not only visualize the space, but also understand and implement the needs of the building owner, taking into consideration construction realities and the site. A qualified designer or architect must have an understanding of building techniques and grasp of materials. A gifted designer or architect blends these considerations into a design that is both practical and stimulating.

The architect or designer creates a written depiction of the design in a set of drawings, or plans. The plans are scaled down drawings, usually using a scale such as 1/8 inch or 1/4 inch. This means that every 1/8 or 1/4 inch on the drawing equals 1 foot in the actual house.

Once the architect or designer has drawn a plan, the plans must be *engineered*—i.e., a structural engineer (not the architect) has to determine how to make the proposed house strong enough to withstand various forces that will be applied to it. The engineer will prepare his or her own set of structural plans that will be used along with the architectural plans to build the structure.

Engineers are concerned with certain forces, or loads, that affect a home: Dead load is the weight of all the materials that are used to build the home. Live load is the weight of people and furniture. Lateral loads are sideways forces applied to a home by wind or earthquakes. Vertical loads are forces from earthquakes that move the house up and down and threaten to tear it away from the foundation.

Engineers also determine how large the pieces of dimensional lumber or metal must be to support the ceilings and roof so that the ceilings and roofs don't sag and so that the windows and doors are not affected negatively by the weight of structures around and above them. When an engineer determines that certain sizes of wood and metal components need to be installed, and that the various components need to be attached together in specific ways, the engineer supplies additional drawings and specifications for the contractor.

Once all of the various drawings and specifications are accumulated into one final set of plans, the plans are taken to the building department of the city, county, or state in which the project will be built. The owner of the proposed project pays a fee to that governing body to have the plans reviewed. At the building department, a building official or plan checker reviews the plans to determine whether they conform to the requirements of building codes, a set of rules adopted by a city, county, or state with certain requirements for the ways in which various buildings are to be constructed. Building codes are written, primarily, by a private organization called the International Code Conference (ICC), the successor to the International Conference of Building Officials, located in Brea, California (the ICBO). The ICC prepares the International Building Code, which is sent to city, county, and state governments. These agencies then adopt the uniform building code as they receive it, or sometimes modify in certain respects. (California adopted the International Building Code for its statewide building code in March 2004.) Uniform plumbing codes and uniform electrical codes accomplish much the same purpose as the Uniform Building Code, which is to give guidance to those in the industry about standards for good construction. One important thing to keep in mind about building codes is that they provide only the minimum standard for construction. The minimum standard may not be appropriate for a luxury home, nor may the minimum standard meet the homeowner's expectations.

The building official or plan checker may require certain changes to the plans, or he or she may request certain additions to the plans. The architect or engineer then makes these changes and once the building official or plan checker is satisfied, the building department will be prepared to issue a building permit.

Why do cities and counties require building permits? First, the building permit process is a way to improve the quality of new buildings, because the plan check process helps to insure that the building will be well designed and will conform to the requirements of the Uniform Building Code, a code that has been heavily researched and developed over many years. Second, the building permit process is a revenue raising device for cities and counties. When the building permit is ready to be issued, the owner is required to pay more money for the actual permit (remember that the owner has already paid a fee to have the plans reviewed). This fee might be substantial, and it might include fees that are intended to pay for parks, schools, and other public improvements. The fee is usually calculated on the square footage of the proposed structure, using a certain rate. For example, if the city's rate for building permits is $1.50 per square foot, and the proposed residence is 5000 square feet, the building permit fee will be $7500. The city may also require other fees to be paid, which can add substantial expense to a project.

Of note in this regard is the issue of school district fees. In some cities, in a remodeling project, the owner is entitled to a refund of at least part of the school district fees. Here is how such a refund works. Assume that the existing house is 5000 square feet. Assume that the new house will be 6000 square feet. Assume that the school district fee is $3.00 per square foot. The city will charge the owner $18,000 for the school district fee. But the owner is entitled to a refund of $15,000 because the existing home (upon which the owner pays property taxes, a portion of which are used to pay for schools) is 5000 square feet. In most municipalities, there is a time limit to apply for the refund, and it is strictly enforced.

Once the building permit has been issued, the construction process is ready to begin. When the actual construction begins, the progress of the construction is monitored by the building department. As noted earlier in the chapter, this is done through a series of inspections. Some of these inspections are performed by an official from the building department, called a field inspector. Some of the inspections are done by people who are paid directly by the owner. Some of these private inspectors are called deputy inspectors, some are consultants or experts.

An example of a consultant or expert would be a soils or geotechnical engineer, who examines the soil conditions before the foundation is constructed or while retaining walls or other structures are being built to determine whether the soil can support these structures. An example of a deputy inspector would be a person who has special expertise in welding. This person would inspect the welds in metal pieces to make sure that the welds have been done correctly so that the pieces will be strong enough to withstand forces that might be applied to them in an earthquake or by wind.

The inspection process requires the contractor to have the municipal field inspector, or the private inspectors, at the job during specified times. For example, the field inspector will want to inspect the foundation before the wood frame is built. The field inspector will then inspect the basic wood frame, the electrical installation, the plumbing installation, the waterproofing, the mechanical (hvac) installation, and the drywall installation. In some cases, such as the placement of structural concrete, or welded metal structures that require special inspections by experts during construction, the municipal field inspector will require reports from the deputy inspectors.

Upon completion and approval of the entire construction, the building department issues a Certificate of Occupancy, sometimes called a "C-of-O." The issuance of a C-of-O is important for many reasons:

• It is evidence that the house has been inspected and that as far as the municipal building inspector is concerned, all of the work has been done in accordance with the plans, specifications, and the building code.

• The issuance of a C-of-O means that the owner can occupy the premises. In most jurisdictions, it is illegal for the owner to live in a new home before the C-of-O is issued.

• The C-of-O is a key date for the contractor, because in most contracts, the contractor cannot receive his or her final payment until the C-of-O is issued.

• When a construction loan has been issued, the construction lender will usually not convert the construction loan to permanent financing (or the homeowner will not be able to obtain permanent financing from another lender) until the C-of-O is issued.

• In many states, the issuance of the C-of-O starts the statute of limitations running for claims of construction defects. This means that if a particular state has a three-year statute of limitations for construction defects, the owner must file a lawsuit within three years of the date of the C-of-O or lose their right to sue.

• It also means that the tax assessor will assess the value of the new home for property tax purposes.

HOW DESIGNERS, ARCHITECTS, ENGINEERS, AND CONSTRUCTION MANAGERS WORK

Architects and designers generally provide their services to homeowners in one of four ways:

• They quote a price that is based on a percentage of the estimated cost of the construction. For architects, this usually ranges from 7 to 15 percent of the construction cost.

• They provide a price on a per square foot basis.

• They work for a fixed fee.

• They work at an hourly rate. Engineers almost always bill on fixed fee or hourly basis.

Some people think that paying an architect a percentage of the estimated cost of construction is a bad idea, because it encourages the architect to make the building more expensive. I do not favor percentage-based architectural fees, and I prefer negotiating an hourly or fixed fee with an architect or designer. I also believe that many times it is more effective to hire a designer, and not an architect, and to have the plans drawn later by a draftsperson. Many young architects are struggling to develop their careers, and these folks have excellent skills in drawing plans and good design ideas. They can draw plans that have been created by a designer and can assist the owner in submitting the plans to the building department, usually for substantially less money than if a more senior architect was hired in the first place. The local architectural schools are a good source for such individuals, who do not yet have established reputations. One caveat: architects, like many creative people, may not focus on the economics of house design, because they are more interested in the esthetic aspects of the project. It is important that you have a budget and communicate your budget to the architect before the design process starts.

A further note about architects. With apologies to the many professional and knowledgeable architects in practice (and there are many), it is unfortunately the case that while some architects have a good design sense, they also have a poor understanding of construction means and methods. This may sound counter-intuitive, but it is true. As a result, many times the architectural plans will be lacking in the necessary details, leaving the contractor to try to figure out how to actually build the building. If the contractor is experienced, this may not be a problem. But an inexperienced or careless contractor using a less than adequate set of plans is a recipe for disaster.

In addition, recall that most construction defect claims start with water intrusion, and most water intrusion claims result from faulty waterproofing of windows and doors and other areas that require great attention to detail. Architects, in my experience, have little familiarity with the specifics of waterproofing. In this particular aspect of the building process, attention to detail means the difference between a dry house and a leaky house. As is discussed in more detail in a later section of this chapter, I recommend the use of waterproofing consultants to avoid the problem of plans that lack waterproofing details.

Construction managers provide a different set of skills. They act as the owner's representative, or the owner's eyes and ears. Construction managers are experienced in the construction industry, and their job is to verify that the building is being constructed in accordance with the plans, specifications, building codes, and best construction practices. Construction managers also review and verify the adequacy of the plans and specifications and can assist owners in the bid process by qualifying contractors and subcontractors. Construction managers also assist the owner in evaluating and approving bills from the contractor and subcontractors and verifying that the appropriate releases have been obtained from suppliers and subcontractors to avoid mechanic's liens.

Most experienced construction professionals agree that although carefully prepared plans, specifications, and details are essential to a successful project, and that attention to detail in the construction process is also important, the most important element of success in a project is supervision. Even the best-intentioned contractor and the most skilled and careful subcontractors make mistakes. Sometimes those mistakes are the result of assumptions that are faulty, sometimes they result from site conditions that do not coincide with the plans, and sometimes they result from simple carelessness. The point is, however, that an experienced construction manager adds another layer or level of supervision to a project, and in this regard, the construction manager can prove to be invaluable to the success of a project.

Although traditionally construction managers were involved only in large commercial or public works projects, or large residential projects, the use of construction managers in more typical residential construction projects has become more common. I recommend the use of construction managers on all but the simplest of projects because the complexity of the construction process requires layers of supervision. In the long run, even though construction managers charge for their services, use of a construction manager will ultimately save the owner money, and in many cases, save the owner substantial sums of money from the start to the completion of the project.

HOW CONTRACTORS WORK

The person who builds or remodels a house as a profession is called a general contractor. General contractors are required to have experience and knowledge in the construction process and are required to be licensed by the state in which they work. However, building a house requires many special and rather narrow skills. For example, a particular contractor might have substantial experience in constructing the wood frame of a house (framing) but little experience in installing the electrical wiring. In fact, to illustrate how narrow the skills are, drywall is a good example. Drywall installation comprises two main aspects: hanging the drywall and finishing the drywall. *Hanging* means attaching the drywall sheets with nails or screws. *Finishing* means applying paper tape and drywall joint compound (called *mud* in the drywall trade) to conceal the seams and the nails and screws. It is common for one crew to hang the drywall and another, totally different crew, to finish the drywall because each job requires a different set of skills.

Typically, the various narrow tasks, like electrical and plumbing work, are performed by people who specialize in just those areas. These people are called *subcontractors*. With this in mind, general contractors work in several different ways to build or remodel houses:

- Most general contractors use subcontractors for nearly every aspect of the construction process, in which case the general contractor's role is to schedule the work and to supervise and coordinate the work to insure that the work is being done properly.

- Some general contractors directly employ in their own company a large number of people who have the experience and training to work on all aspects of the construction process, meaning that they build every aspect of the house and do not use subcontractors.

- Some general contractors use a mix of these, doing some of the work themselves and using subcontractors for other work.

Because the construction of a home involves the work of many specialized people, or trades, the scheduling and coordination of the work as well as the inspection process is an important part of the general contractor's role in the process. One example is the installation of electrical wiring. The electrical subcontractor cannot begin to install the wiring until the wood frame of the house has been built, because the electrical wires and the conduit containing them are run through small holes cut into the wood frame. Each of the trades must be scheduled so that they are available to work at the right time and so that their work does not interfere with the work of other trades. A typical and important cause of delay in construction is the failure to schedule and coordinate the various trades.

How Contractors Charge for Their Work

Contractors charge for their work in one of the following ways:

- By using a fixed, or flat, fee for the construction, sometimes called a *stipulated sum* or *guaranteed maximum price* (GMAX)
- By charging the owner the actual cost of the materials and the labor costs, at labor rates agreed upon at the outset, plus additional charges for overhead and profit, sometimes called a *cost plus* contract
- By charging a fee for supervision only, when the owner contracts directly with subcontractors for their work
- By charging a fee on a *not to exceed* basis, which is basically a fixed fee that is reduced if the contractor's actual costs are less than the contractor estimates.

The terms *overhead* and *profit* are used many times in contract discussions. A contractor's overhead is simply his or her fixed costs of operation. For example, if a contractor has a large operation, there will be, among other things, office rent, telephones, payroll, insurance, and vehicles. The contractor takes his or her total overhead and tries to allocate some of that overhead to a particular job. An additional item, often misunderstood, is the cost of general conditions, or field conditions. General conditions usually include the following: A temporary office (such as a trailer or movable shack), portable toilets, a job-site telephone and fax, temporary fencing, a temporary power pole, temporary water hook-up or supply, and other security measures (lately, security cameras linked to an Internet connection that allow 24-hour surveillance of the site).

General conditions are not part of the contractor's overhead. General conditions are charged separately to the owner as an additional cost of construction. Similarly, field staff, such as a project superintendent or project supervisor, are not part of the general contractor's general overhead and they are charged separately as well, unless an agreement stipulates that supervision will be part of the contractor's overhead and profit.

On complicated jobs, where many people are working and many deliveries of goods and materials are made, it is important that someone keep track of what happens on the job every day. To effectuate this, the contractor uses an individual called a *clerk of the works*, who records the workers, the deliveries, and various other items of information on a daily basis. The cost of this individual is charged as a job cost, and not as part of the job overhead. As a side note, even if there is no official clerk of the works, keeping a record of deliveries of goods and materials delivered to the job is an essential part of the overall record-keeping function of a general contractor.

Because it is difficult to allocate general overhead accurately to a particular job, over the years the industry has evolved to the point where most contractors simply use an arbitrary figure for both overhead and profit. These figures vary considerably, but the important thing to know is that they are subject to negotiation between the parties. A typical range would be 12 to 20 percent combined overhead and profit. Contractors who use the lower range of numbers usually combine the numbers; contractors who charge at the higher end usually split the numbers into 10 percent overhead and 10 percent profit. The fact is that the allocation is irrelevant to the owner, because the owner is going to pay 20 percent in that example, and the owner doesn't particularly care how the contractor allocates the number internally.

What is important is the ultimate percentage charged by the contractor. Many people think that a combined number of 20 percent for overhead and profit is standard in the business. This is not true. There is no particular standard figure for overhead and profit; it is purely up to the owner and the contractor. The percentage is more a function of shifting risk, as discussed in more detail in the following section.

WHAT IS THE BEST KIND OF CONTRACT?

There is no perfect answer to this question, and there is no perfect way to structure the economic relationship between the owner and the contractor. A construction contract is a device that, from the economic perspective, serves to assign and allocate risk. Risk can never be allocated perfectly in construction, which means that in every project, either the owner or the contractor will lose some amount of money that in a perfect economic structure they would not have lost. Understanding this basic fact will go a long way in making you a sophisticated consumer of construction services and will help you understand the negotiating process.

Before discussing the economics of a construction contract, an introductory word about the purpose of the contract itself is important. The details of the contract are discussed at some length in Chapter 2. For purposes of overview, however, the contract basically serves only a limited number of functions and has two major elements: price and performance. A good construction contract will ensure that there is no confusion about the following:

- What the contractor is supposed to build
- How the contractor is supposed to build it
- How much the work is supposed to cost
- How long the work is supposed to take
- Who is going to do the work
- What quality of work is to be provided
- What mechanism will be provided so that the contractor and subcontractors are paid for the work they do but are not overpaid
- What mechanism will be provided so that the subcontractors do not file liens against the property

- What happens if the cost goes up or down
- What happens if there are delays in the work
- Who pays for changes in the work and how are changes to be charged
- How the owner will receive the quality of work and the timeliness of performance that he or she bargained for
- How the contractor is to receive clarification with regard to questions about the plans and specifications (usually through a form called a request for information, or RFI)
- What insurance will be provided

The economics of construction contracts, however, require a much more detailed and sophisticated analysis. Construction, like virtually all business ventures, involves risk. The risk arises from the fact that neither the owner nor the contractor know precisely in advance what it will cost or how long it will take to complete the job. Both the owner and contractor are limited to estimates for the cost and time to complete the job. In addition, because the contractor is not paid for the entire job in advance, the contractor is required to "front" at least some money for the labor and materials. If the contractor has underestimated the cost, or if other factors result in a delay in the contractor getting paid, the contractor might run out of money to finish the job or might finish the job but not make any profit.

The most important thing to understand is that the economic relationship between a contractor and an owner is based on an allocation of the inherent risk in the construction process. The construction contract is, in some ways, nothing more than a balancing act, where both parties try to balance the risk and minimize their exposure. Because of this balancing act, an owner who is able or willing to bear more of the risk can generally make a better bargain. Conversely, an owner who wants to shift more of the risk to the contractor will have to pay for that privilege.

An example of a typical contract will help to illustrate this idea. When a contractor proposes a fixed-fee agreement, both the owner and the contractor assume some risk. The contractor assumes a risk that his costs for material and labor might be more than his estimate. If this occurs, the contractor will make a smaller profit or perhaps no profit on the job. The owner assumes a risk that the contractor's estimate will be more than the actual cost of the construction. If that occurs, the owner will have overpaid, because the contractor's profit will be much more than the owner might think is reasonable.

The parties attempt to minimize, or appropriately allocate, this risk by several means. First, there are market forces at work. A smart owner will require multiple bids for the job. Multiple, in this context, means at least three bids from *qualified contractors*—contractors who have the experience and labor force necessary to build the project, who have reviewed and analyzed a complete set of plans, and who have bid the job by getting bids from qualified subcontractors (in other words, the contractor is not "guesstimating" the cost of the project).

A contractor whose charges are too high will likely not get the job. Because of the market forces, the contractor tries to estimate the costs as closely as possible. To do this, he or she presses the designer, owner, and engineers for details of the construction, so that the contractor doesn't have any surprises down the road. A good contractor pays particular attention to the expensive components of a home, which are the finish items such as flooring, appliances, plumbing fixtures, electrical fixtures, and hardware, because the cost of these items can vary enormously.

All of the various ways to structure the economics of a contract therefore have certain advantages and disadvantages. The pros and cons of a fixed-price contract have been discussed. The advantage of a cost and materials contract, sometimes called a cost plus contract, also has certain advantages and certain disadvantages.

The advantage of a cost plus contract (the cost of labor and materials plus overhead and profit) is that the owner pays only the actual cost of materials and an agreed-upon labor rate, plus overhead and profit. Thus, the owner has less risk of overpaying the contractor—i.e., giving the contractor too much profit on the job. A disadvantage, however, is that a "time and materials" job can sometimes serve as an encouragement to make the job bigger, not smaller. For example, if a particular aspect of the job involves the construction of custom cabinets, the contractor could stretch out the time "necessary" to build these cabinets and make more money on the labor. A cabinet subcontractor who bids the construction and installation of the cabinets on a fixed-price basis has an incentive to build and install the cabinets quickly, because every hour that his installer is on the job, he is losing potential profit. (Remember that if a fixed price contract were used, the cabinet subcontractor has profit built into the job if he or she can do the work quickly, and he or she will make more money but bears the risk of losing money if it takes longer than expected.)

Another disadvantage of the cost and materials approach is that it is difficult for an owner to know how long it should take to do the work. For example, the contractor might have to build a wood form to pour concrete for front steps, arrange for the concrete to be delivered, pour the concrete, and finish it by stripping away the forms after the concrete is poured. Let's say that the contractor spends 30 hours doing this job and charges $25.00 per hour for the labor. It is entirely possible that the same job could have been done in only half the time, in which case the owner has overpaid by a considerable sum. Few owners would know whether that particular job should have taken 30 hours or 15 hours. And, in contrast to the fixed-price bid process, where the owner would get three or four bids for the work, there is little ability for the owner to "test" the validity of the charges.

One way for an owner to protect himself or herself is to obtain bids from the major subcontractors on a fixed-price basis. These major subcontractors would usually include the foundation contractor, the framing contractor, the plumbing contractor, the hvac contractor, the electrical contractor, the drywall contractor, the roofing contractor, and the stucco contractor (assuming a stucco finish on the exterior). Three bids for each of these trades will give the owner an excellent basis for knowing whether the prices being charged are realistic.

In many instances, because of the risk allocation concept that forms the basis of all construction contracts, the contractor simply will not or cannot agree to a fixed price. This happens often in remodeling projects because the contractor is concerned about finding conditions that might take a great deal of time and effort to correct, and the contractor doesn't want to bear the economic risk of those unknown conditions. In those circumstances, a time and materials approach is often the only solution, or at least the only practical solution because a fixed-price contract would be too expensive (so that the contractor minimizes the risk).

One alternative in that situation would be to spend some time and money investigating the conditions before negotiating a contract. This will require some degree of "destructive testing," meaning an intrusive investigation at the site. Destructive testing might involve the removal of drywall on the interior, or stucco or other exterior finishes on the exterior, to determine the condition of the foundation, the framing, the electrical, and the plumbing installations. Oftentimes, even though this process is messy and can be expensive, it will save money over the long term because both the owner and the contractor will have better information upon which to base their discussions.

To a large extent, the whole issue of price also depends on a number of factors that are external to the contract. For example, the contract price depends heavily on the contractor's experience and ability to estimate costs. A contractor who is not familiar with the particular type of construction, or who has failed to review the plans and specifications carefully, might prepare a bid that is substantially inaccurate. If the bid is too high, the owner might overpay for the job. If the bid is too low, the contractor will lose money, and that job will be headed for a great deal of trouble.

By the same token, even a contractor who is able at the estimating process can sometimes be less than honest with the owner. The contractor might attempt to take advantage of a naive or unsophisticated owner and overcharge. The contractor might also do the opposite: undercharge, because the contractor wants to get something other than a profit from the particular job in question. Some examples would be a contractor who wants to create a relationship with the owner because the contractor perceives that the owner will bring him or her other, more profitable jobs from friends or business associates. Or, a contractor may want to start building in a particular location because of a perceived opportunity for more business. If the contractor has the economic wherewithal to finish the job and not make money, the job might turn out well, but jobs for which the contractor makes no money, or loses money, almost always turn out poorly.

Finally, on this subject, a great deal of the contract price and the contractor's performance depends on the contractor's integrity. If the contractor is honest and has a high degree of integrity, the contract price will be fair and the contractor's charges (on a time and material basis) will also be fair. Contractors who recognize that every good job and every satisfied owner will lead them to other good jobs are highly motivated to do good work at a fair price. Those who look at every job as an opportunity to maximize their profit at the expense of the owner will not have much of a future in the business.

There are ways to "check out" a contractor. A good contractor will have references, and, by all means, the prospective client should call each of the references. A prospective client can ask the contractor for the names of subcontractors with whom the contractor has worked in the past. The prospective client should call each one of those subcontractors, and ask the following:

- Were prior jobs completed on time and on budget?
- Were there any problems in the subcontractors receiving timely and complete payments?
- Were the jobs organized, or were there delays because of a lack of coordination between the trades?
- Were the owners satisfied?
- Has the contractor ever been involved in litigation over money or quality issues?

In addition, the various state agencies that license and regulate contractors maintain publicly accessible databases of information on contractor discipline and complaints (See, for example, the California Contractor's State License Board website: www.cslb.org.). The licensing agency's public records will also confirm that the contractor is licensed and that the license is current, and it will provide information about the contractor's bond, all of which should be confirmed by the owner before signing a contract.

In all cases, the owner should insist upon seeing a copy of the contractor's licence and the owner should understand clearly what kind of business entity the contractor uses to conduct business. Most contractor business entities have no assets, perhaps only a few tools or a truck. Thus, it becomes critically important that the contractor have adequate insurance so that there is a source of funds from which the owner could recover damages in the event of a construction defect lawsuit.

Lastly, the contractor must have the financial ability to run the job. Even when the contractor receives a deposit, there is always a gap between the time that the contractor incurs expenses and the time the contractor is paid. If the contractor does not have the financial ability to bridge this gap, to keep payments flowing to his workers and to suppliers and to subcontractors, even the most honest contractor with the highest degree of integrity will not succeed in finishing the project.

2 The Construction Contract

FOR OWNERS AND CONTRACTORS

Let me start with a disclaimer. A truly comprehensive discussion of construction contracts is beyond the scope of this book. One reason is that many contractors (and virtually all architects) use a standard form of contract, developed by the American Institute of Architects (AIA), and the various provisions of the standard AIA contracts have been the subject of intense scrutiny and court decisions for many years. As a result, hundreds of articles, books, court decisions, and commentaries have focused on the AIA contract. Numerous books and articles on construction contracts in general have been written by experts in the field. In addition, many contractors use their own form of contract or contracts prepared by trade groups, such as the Associated General Contractors (AGC).

Each of these "form" contracts have a built-in bias. The AIA form, in my opinion, is biased toward architects and gives them more protection than perhaps they are entitled to. The AGC form, not surprisingly, is biased toward the contractor and against the owner. In addition, neither contract deals particularly well with the issue of how to resolve disputes during and after the construction process, and neither contract provides any meaningful assistance to the central issue of avoiding construction defects.

My purpose in discussing construction contracts is the same as my overall purpose for writing this book: to help owners, construction professionals, and developers avoid the sort of mistakes that seem to occur repeatedly. For this reason, I will focus only on the essential terms of construction contracts terms that seem to be lacking in many contracts, whether using the AIA form or another.

The contract serves three basic purposes: (1) to make it clear what is to be done, (2) to make it clear how long it will take, and (3) to show how much it will cost. Although it can take a lot of words to cover those items, a good construction contract will leave the parties with no doubt about the exact price that is to be paid,exactly what work will be done, what quality of work will be done, and how long it will take to do the work.

COST OF CONSTRUCTION CONTRACT PROVISIONS

In discussing contract terms, I like to start with the objective of the contract and work backward. The objective of the price part of the construction contract is to make sure that the parties know exactly what the price is and what the price is intended to cover.

In a *fixed-price contract*, sometimes called a *stipulated sum contract*, the total amount of the price for the job is obvious, but what the price is supposed to cover sometimes isn't so clear. This is particularly true in two instances: (1) where the contract provides for allowances, and (2) where the contract provides for change orders. Let's talk about allowances first.

An *allowance* is an estimate for the cost of a particular component—such as a marble counter-top for a bathroom vanity. To get the job started and sign a contract, the parties provide for an allowance for the marble top. In many instances, the parties cannot determine in advance the exact prices for certain aspects of the work.

In this case, the actual cost of the marble top can vary. It could be more or it could be less than the allowance in the contract. The contract must therefore specify what happens in either case. A good contract will specifically state that if the actual cost of an allowance item is less than the number set forth in the allowance, the difference will be deducted from the contract price—i.e., it is returned to the owner. If the amount is higher, the owner will have the choice of paying the additional amount or specifying a different material or component that might be closer to the allowance in the contract.

One good way to avoid potential disputes on allowance items is to spend some time researching the cost of these items, eliminate as many allowance items as possible from the contract, and as to the remainder, have a good sense of what the likely price difference might be.

With regard to *change orders*, a slightly different issue is involved. First, let's take a moment to explain the concept of change orders. Changes during construction are a fact of life. Unexpected things occur, owners change their minds about the design, or as the project evolves it becomes obvious that certain things need to be changed. Sometimes the building official makes a decision that requires a change. Change orders are a mechanism to adjust the price and time of the contract to take these changes into account. Technically speaking, a change order is initiated by the contractor in response to a changed circumstance in the job. When the owner wants to make a change, the owner sends the contractor a construction *change directive*. The two terms are often confused, but a good construction contract will differentiate them because the cost of a change order can vary from the cost of a construction change directive.

A good construction contract will, however, provide that the contractor is to obtain a signed (by the owner) change order before making a change to the work and that the contractor will sign all construction change directives from the owner before starting the work specified in that document. The change order and construction change directive will specify the nature and extent of the change and will state whether the contract price will be increased or decreased as a result of the change, and whether the time to complete the job will be made longer or shorter. A good contractor will also make sure that the change order documents any other impact from the change (for example, if a lesser quality material is substituted for cost reasons, a good contractor will note that the maintenance of this item might be more costly or that its useful life may be shorter than a similar component of better quality).

When the parties can easily agree on the cost of the change, no problems result. Problems arise when the cost of the change isn't so clear. When this occurs, the contractor will usually want to charge on a time and materials basis. To avoid disputes, the contract needs to specify the labor rate, the overhead and profit, the increased cost of general conditions, and the increased cost of insurance, and the contractor will need to provide a mechanism for dispute resolution in the event that the parties cannot agree on the time to be spent for the change.

A form of change order (and there are many forms of such documents) can be seen on the next page.

SCOPE OF THE WORK CONTRACT TERMS

The question of what work the contractor will do is called the *scope of work*. Scope of work issues can arise in a number of contexts. In new construction (meaning the construction of a brand new home from the ground up, as opposed to remodeling an existing home or building), the issue is usually whether the owner wishes to have some of the work done by someone other than the general contractor. An example would be an owner who wants to have an artist apply a special finish to the walls instead of a more conventional paint or wallpapered surface. Another example would be an owner that wants to contract directly to have some aspect of the work done by a specialty subcontractor, such as a landscaping contractor or a swimming pool contractor.

Generally speaking, scope of work issues cover three considerations: The owner and contractor agree on what the contractor will do. The owner and contractor agree on the impact that the outside subcontractor might have on the job. The issue of liability is considered—meaning, how the owner's subcontractors will impact on the contractor's work and the contractor's liability.

Agreeing on the contractor's scope of work is not difficult. It requires a clear understanding of what the contractor is including in the job. One good way to avoid disputes on scope of work is to imagine what the project will look like when it is complete. Start your thinking from the approach to the house or building. Is the driveway and curb work included? Are the landscaping and hardscape (concrete, brick and stone, concrete steps, driveways, and walkways) included? Are the exterior light fixtures included? If the utility connections are to be buried between the utility line and the home, is the cost of burying them included? If a pool or outdoor patios exist and require electrical, are the electrical lines to that equipment included in the contract?

General Contract—Change in the Scope of the Work

ABC Contracting Company 124 Concrete Avenue, Buildertown USA.

Date _____

Change Order No. _____

To Owner re _____(Project Location):

The General Contract shall be revised pursuant to the following changes in the scope of work:

Request for Information No. ____ (if applicable) _____

Date Response received: _____

The following change(s) in the Scope of Work have been requested by and are accepted by the Owner:

_____

_____

_____

_____

_____

Subtotal: $

General Conditions: $

Overhead __% $

Profit __% $

Insurance __ % $

Total of this Change Order shall not exceed: $

The Contract Agreement Time shall be increased _____ or decreased _____ by _____ days and

therefore the date of Substantial Completion shall be _____.

ABC Contracting Company

By:

_____

Its _____

Approved By

_____

Owner or Owner's Authorized Agent

Continue your imaginary tour inside the home. Visualize every room, starting at the floor, going up to the ceiling, and including everything that you can see and feel. Are the interior light fixtures included? Are the plumbing fixtures included? Are the appliances included in the contract price? Are the wall-coverings and floor coverings included? What about telephone, computer, and sound systems? Is a security system included? Think about other finish items, such as millwork (things like moldings, built-in cabinets, and paneling), bathroom vanities, or fireplace mantels and hearths. Are these included in your contract?

Try to visualize every room slowly and carefully, as if you were living in the house. If you put out your hand expecting to find a curtain or window shade to cover a window, will it be there or will you have to pay someone else to install it? If you have a pantry, will it have shelves, and will the contractor charge extra for them? Will the closets have shelves, and will the contractor charge extra for them? Will the light switches and receptacles be standard or more expensive custom items?

A slow, methodical, and careful review, followed by a careful discussion with the contractor will avoid disputes later on the scope of work. In addition, the amount of detail in the plans has a direct relationship to minimizing scope of work disputes and problems. If the plans contain specific details, everyone involved can see those details and the parties can discuss and agree upon who should be responsible for the installation of the items shown.

The second scope of work issue concerns delays and disruptions to the job caused by people working on the job who are not employed by or answerable to the general contractor. Let's say that you have a friend who is a plumbing subcontractor. Your plumbing subcontractor friend tells you that she can do your plumbing work for considerably less money than the general contractor's plumbing subcontractor.

You approach the general contractor and tell him that you want to remove, or delete, the plumbing work from the general contractor's scope of work. Although you might in fact save some money on the plumbing work, you could be opening the door to a host of problems. Here are some examples:

- The plumber promises to be done by a certain date. The contractor schedules another subcontractor to follow the plumbing work. The plumber is late, and the subcontractor who is supposed to follow her cannot start on time. That second subcontractor then has to take another job, and cannot begin his work on your project until the other job is done. The subcontractors that are to follow that subcontractor are also delayed, and the entire job is thrown into disorder.

- The plumber does poor work and a water line breaks or begins to leak, and floods the job. The water damages other components of the construction. The contractor takes no responsibility because you and not he employed the plumber. The job stops while everyone tries to resolve the problem.

- Without being aware of it, the plumber cuts through an electrical line installed by the electrical subcontractor. Later, when the walls are closed up, the electrical circuit doesn't work. Thousands of dollars are spent and hours of time expended in finding the problem. The general contractor refuses to pay for the problem, claiming that your subcontractor, and not his, did the damage.

These are only a few examples of what can occur when the overall scope of work is divided between the general contractor and others who don't work for the general contractor or who are not legally responsible to the general contractor. A common additional example is that the owner's subcontractor simply works too slowly, or he or she gets in the way of other subcontractors, leading to a slowdown in the overall job.

A well-written construction contract will specify what is to occur when the owner and contractor divide the scope of work, and it will clearly define the responsibilities of the owner and the contractor in the event of delay, damage, and other disruptions to the job.

QUALITY OF THE WORK CONTRACT TERMS

Every owner has his or her own set of expectations for the quality of work on the job. Sometimes these expectations are clear to the owner. Sometimes the owner has only a vague set of expectations. Similarly, every contractor approaches a job with a set of expectations of what that job will "look and feel" like when it is complete. The trick is to make sure that the owner and the contractor share the same set of expectations and that the contract makes those expectations clear.

Let's start with a few basics that should be in every construction contract, no matter how elaborate or simple the job. Those basics include five elements: (1) that the work will follow the plans and specifications; (2) that the work will comply with all applicable governmental regulations and requirements; (3) that the work and materials will conform to the recommendations of the manufacturer of the product; (4) that the work will be of a quality not less than the quality of work done by others in the community for the level of fit and finish for the particular project; and (5) that no materials or methods of inferior quality will be substituted without the consent of the owner or the owner's representative.

Consider these one by one to get a better understanding of why they are important.

Contract Terms Requiring the Contractor to Follow the Plans and Specifications

Remember that every building project starts with plans and specifications, and those plans and specifications are reviewed by the building department and approved before the building permit is issued. The simplest place to start, therefore, is to say that unless the project is built according to the plans and specifications, it may not comply with the building code and may not be constructed properly. If deviations from the plans and specifications occur, the building may be structurally unsound.

At a different level, because the contract will require that the contractor build the building in accordance with the plans and specifications, if the general contractor fails to do so, the general contractor has breached the contract. The general contractor may then be legally responsible for the cost to make the project conform to the plans and specifications, and for other consequential damages, such as the cost to the owner to move out while remedial work is being done. I like to specifically reference the plans and specifications in the construction contract, and then have the general contractor initial each page of the plans and specifications. I also like the construction contract to contain language that states that the contractor has read every page of the plans and specifications, that the plans and specifications are adequate to construct the job, and that all the plans and specifications will be read by all of the subcontractors. Additionally, it is important to include a provision in the contract that states that the plans and specifications are complementary to one another and are to be read together.

Contract Terms Requiring the Contractor to Comply with Governmental Regulations

This requirement is related to, but not exactly the same, as building the structure in compliance with the plans and specifications. An example will illustrate the difference and the importance of this provision. Suppose that you are building a home on a hillside and that a certain amount of grading (moving of earth) is involved. In that situation, the building department will likely ask the grading department of the city to establish specific requirements for your project. The city will likely also ask you to post a grading bond, a deposit of money, to insure that the project complies with all the city's requirements. Suppose also that one of those requirements is that the contractor establish a plan to mitigate the effects of rain on the property during construction, or to provide for methods to control erosion from rain. Such a plan could consist of sandbags, plastic sheeting, or temporary drains.

Now suppose that your contractor doesn't do this, and substantial rainfall occurs. Mud from your property flows onto the property of the neighbor or into the city street. Perhaps other parts of your construction project are damaged. The city sends its own crews in (at an enormous expense) and deducts the cost from your bond.

If you have a provision in your contract that mandates that the contractor will comply not only with the plans and specifications, but also the requirements of the governmental authority, you will have recourse against the contractor for failing to comply with the regulations.

In California, construction defects are defined by statute (California Civil Code section 896). Referencing this statute in a construction contract by stating that any violation of the statutorily defined construction defects will be conclusive evidence of construction defects (and a breach of the construction contract) would be of substantial assistance in making it clear to all parties what is expected under the contract.

Sometimes the governmental agency, acting through the building inspector, imposes additional requirements during construction that neither the owner nor the contractor expected. These changes usually result in additional cost for the project. Unless the contractor should have anticipated these changes in advance, the owner is likely have to pay the extra charge.

Complying with Manufacturer's Recommendations

The construction of a new home, or the remodeling of an existing home, requires the use of hundreds of manufactured products. Some examples are waterproofing materials, sealants, insulation, windows, doors, electrical cable, adhesives, hardware, plumbing fixtures, and roofing materials. It is common for construction contracts to specify the use of particular materials because the owner may want a certain standard, or a consultant may believe that the product in question represents the best solution for a particular construction problem.

A contract requirement that specifies that the contractor will follow the manufacturer's recommendations is important for two reasons: First, even the best product will not perform if it is not installed or applied correctly. Second, the manufacturer will not provide warranty protection for products that are not installed or applied in accordance with the manufacturer's recommendations.

Here is an example. When windows are installed in the openings in the wood frame, the gap between the window frame and the opening must be sealed so that water does not intrude into the inside of the home or into the wood structure. Water intrusion will cause dry rot and has the potential to cause mold and mildew infestations inside the home. Various types of material, generically called *flashing*, are installed around the window frame to prevent the transmission of water through the opening. A popular type of flashing with proven reliability works only if it is applied correctly. The application process specified by the manufacturer requires the window frame to be primed, so that the flashing adheres firmly to the window frame. If the primer isn't applied, the material won't stick: it's that simple. No matter how good a product, it will be totally useless unless the contractor strictly complies with the manufacturer's recommendations.

Here is another example. When fiberglass spa tubs are used, the weight of the water can be considerable. The water weight can cause the bottom of the tub to flex. Over time, the flexing will cause the tub to crack. Some tubs have a "foot" integrated into the bottom of the tub during the manufacturing process to provide support so that it resists flexing. If the tub does not come with a factory integrated foot, the manufacturer will usually specify that a supporting material be installed under the tub (commonly a small mound of mortar or similar material). Without knowing and following the manufacturers instructions, the tub could be installed in the wrong fashion and fail over time.

Community Standards

The shorthand version of constructing a building in the way that others in the community would build a building of the same general quality is called *standard of care*. The standard of care is an important legal term in any case involving a claim of construction defects. To prove that a construction defect is present, the plaintiff must show that the contractor violated the building code, failed to install a product in conformance with the manufacturer's recommendations, or fell below the standard of care. Standard of care means that the contractor must use construction means and methods that are the same as those used by others in the same locality under generally the same circumstances. Some examples will help you understand what this means.

In parts of the world where winters bring considerable amounts of ice and snow, special provisions must be made for the construction of roofs so that roof structures are not damaged by a buildup of ice and snow. Because every contractor building in that location would make provision of some type to protect roof structures from ice and snow buildup, a contractor who failed to take any precautions would fall below the standard of care. Similarly, in areas where tornadoes or hurricanes require the installation of special devices to secure roofs to homes, a contractor who failed to install such devices (even if the plans did not specify them) would have acted below the standard of care.

The standard of care is proved in court by the use of expert testimony. Usually, the plaintiff uses a general contractor who has been building in the same area for a considerable period of time to testify what standards and practices are used in that area.

I like to specify that the contractor will at least meet the standard of care, because I feel that having an explicit contract term encourages the contractor to strive for better quality and also removes any doubt as to what needs to be proved in court. If the plaintiff proves that the contractor fell below the standard of care, the plaintiff arguably has two legal avenues of recovery against the contractor: one theory based on negligence and one theory based on breach of contract, because the contract specifies that the contractor will build to the standard of care. What's the reason for two theories? Without a contract provision, the plaintiff would have only the negligence, or tort, theory of recovery against the contractor. In that event, the plaintiff might win but would not recover his or her attorneys fees. If the standard of care is included in the contract, and the contract also provides that the winning party recovers attorney's fees, the plaintiff would recover the attorney's fees in addition to the cost of repairing the damages.

Note that the standard of care can vary according to the nature of the home that is being built. A luxury home requires a higher standard of care than a tract home (although both require some minimal standard of care). If you expect a higher standard of care, your construction contract should say so.

Substitution of Materials

It is common for a contractor, particularly on a fixed price contract, to look for materials that might do the job as well but cost a little less money. In many instances, this does no real harm, but sometimes it can present substantial problems.

A recent example from a construction defect case shows what can happen when the contractor decides to make substitutions without the permission of the owner or the architect. In this particular case, the owner was building a large and elaborate home with a great deal of woodwork on the exterior of the house. In various areas, the plans called for large, smooth, wood panels, surrounded by other decorative wood elements. The plans specified a specific product manufactured by a specific company. This product has a smooth and uniform surface and is intended to be weather-resistant in exterior applications. The builder substituted a type of exterior grade plywood. The surface of the plywood was not uniform or smooth, and it would have taken a great deal of hand labor from the painter to make the surface acceptable. Worse, before the material was painted, it got wet and started to deteriorate, even though it can be used outdoors. Eventually, all of the panels had to be removed and replaced with the product that was originally specified, a process that slowed down the completion of the job and interfered with other work on the project.

A contract provision that mandates that no material will be substituted without the prior written consent of the owner or architect will prevent these problems. Note that the provision requires *written consent*. In this respect, this particular contract provision mirrors a more general practice that is essential to building and remodeling without conflict and controversy, and that is the rule that *everything should be written and signed*. There is, simply put, no better way to avoid conflict later on in the process than having important issues set forth in writing and signed by all parties.

Time of Completion

The completion date of a project is critically important to both the owner and the contractor. In every single project, the owner is anxious to get back into the house. The contractor is anxious to get paid and move on to the next job. After construction defects, delay in the completion of a construction project is the single most common cause of litigation in the construction field.

As mentioned earlier, much of the construction process is about sharing or allocating risk. In this respect, completion time is no different: it also involves an allocation of risk.

At the most fundamental level, completion time is affected by the following:

- Problems that are caused by the owner
- Problems that are caused by the contractor
- Problems that are caused by neither the owner nor the contractor, but for which the contractor should be responsible
- Problems that are caused by neither the owner nor the contractor, but for which the contractor should not be responsible

Here are some examples of each category:

- The contractor encounters unanticipated site conditions (for example, the job requires excavating for a basement and the contractor encounters rock or water where earth was expected). Whether the owner or the contractor bears responsibility for this condition depends on a variety of factors. The contract could state that the contractor is to bear the liability for unanticipated site conditions, in which case the contractor must inspect the site carefully to insure that he or she isn't surprised once the job begins. If it is impossible to determine what is under the surface of the ground, the contractor may not be willing to take the risk of unanticipated site conditions and the risk for them may shift to the owner.
- The weather changes shortly before or during the job and the job cannot progress until the project is dry or warm enough to continue. In this case, the contractor is usually not responsible for the delay, although the contract should contain provisions for exactly how long weather-related delays are to extend the contract time.

- A worker's strike affects the shipment of materials. In this case, the contractor will ordinarily not be penalized for the delay because the cause of the delay is out of the contractor's hands and could not have been anticipated or prevented using good care and diligence.

- A key subcontractor goes out of business and it is difficult to find a replacement. In this case, the contractor should be responsible for the delay, even though "technically" the contractor had nothing to do with his subcontractor going out of business. The theory behind the contractor having responsibility in this situation is that the contractor should know, in the exercise of due care, whether a subcontractor is likely to remain in business. If the contractor chooses a particular subcontractor (perhaps to save money on a fixed price contract) and that subcontractor goes out of business, the contractor should bear the loss and not the owner.

- The owner makes changes to the project during the construction process (probably the leading cause of delay). In this situation, the owner should bear the responsibility for delays, because the owner has changed the scope of the project.

- The owner runs out of money. This is an owner-responsibility situation.

- The contractor runs out of money (this is a another important reason why jobs slow down and is discussed extensively a bit later). When a contractor cannot complete a job because he or she has run out of money, the contractor must bear the burden of any delay or loss to the owner.

- A subcontractor runs out of money (usually for the same reason that a contractor runs out of money). This is a contractor problem, and the contractor should bear the burden of delay. When the contractor chooses the subcontractor, the contractor bears the risk of making a poor choice or not managing that subcontractor effectively while the job is progressing.

- The plans are inadequate. This is a difficult issue. A good contract will require that the contractor confirm that he or she has read the plans thoroughly and that the plans are adequate to complete the project. It is my view in this situation that the contractor should bear the responsibility for delay in the event that the plans are not adequate to complete the job. A qualified contractor is supposed to be able to read and understand plans: that is one of the basic things that an owner pays a contractor to do.

- The building inspector insists on changes during the job. Unless the building inspector has asked for changes because the contractor is doing a poor job, the contractor should not be responsible for delays caused by building inspector requested changes.

- A structural failure or other catastrophe occurs during construction, such as a fire or flood. Unless the contractor is responsible for the flood or damage, the contractor will not bear the burden of delay from these causes.

- The owner or architect determines that the contractor has not complied with the plans and specifications and the contractor is forced to demolish portions of construction and rebuild the project to conform to the plans and specifications. This is clearly a contractor caused problem and the contractor must bear the loss from the delay.

Contract Provisions Dealing with Insurance

This is one of the most important, least understood, and most difficult aspects of construction contracts. Insurance coverage in construction projects is another example of allocating risk. To allocate risk, the risk must first be identified and a decision made about who will bear that risk. A number of categories of risk must be considered in the construction process. Not surprisingly, as the complexity of the project increases, so does the complexity of the risk. Even in a simple project, risks are present. Here are some of the principal areas of risk:

- People or property could be injured during construction.
- The owner might lose the use of the property through damage or delay.
- Damage could occur through defective workmanship or materials.

Several basic kinds of insurance are used in construction. Each is intended to cover different types of losses. The most important types of insurance policies are these:

- Liability insurance
- Worker's Compensation insurance
- "Builder's risk" policies
- Professional liability insurance policies

As construction defects have become more pervasive, and as insurance companies have paid out hundreds of millions of dollars in construction defect cases, insurance coverage has become more difficult to obtain and much more expensive. In 2005, for example, a commercial general liability policy for a 16-unit condominium project in California would likely cost in the range of $600,000. In many instances, depending on the loss history of the contractor (how many times the contractor has been sued and what payments have been made on his or her behalf), it is probable that the insurance company would not write the policy even if the contractor were willing to pay the premium.

In part as a reaction to the expense of conventional policies, the insurance industry has begun to offer different kinds of products. The newer products are generically referred to as *wrap policies*. More specifically, they are sometimes called *OCIPs* (*owner controlled insurance policies*). In this type of policy, the insurer sells a policy to the owner or the contractor and the policy covers everyone concerned (the owner, the contractor, and the subcontractors). If the policy is sold to the owner, the contractor and the subcontractors do not bring their own separate insurance policies to the project. However, the contractor and the subcontractors sometimes pay a proportionate share of the premium in exchange for receiving coverage under the policies.

More than a few unknowns and considerations are apparent with respect to wrap and OCIPs resulting from the fact that not enough substantial experience with these policies exists in the marketplace and in the courts. Two of these issues are indemnities and costs. On the issue of indemnities, in the traditional model of construction contracts, the general contractor agrees to indemnify the owner and the subcontractors agree to indemnify the general contractor. Suppose that a lawsuit is filed to obtain damages for construction defects. The wrap or OCIP mandates that one lawyer represent all the parties and that the parties work together to provide a "common defense" to the owner's claims. However, in the real world, if there is a construction defect, someone (or some combination of people) is responsible for that defect. The person or persons who caused the defects must ultimately be made to indemnify the people who did not cause the defects. It is probable that the issue of indemnity will have to be determined in a separate but related proceeding after the initial claim by the owner is resolved.

The issue of cost is somewhat simpler. Again, in the traditional model of construction insurance, the general contractor and the subcontractors provide their own insurance at their own expense (and presumably pass through this expense to the owner in the form of their contract price or sometimes as a separate line item). In the case of an OCIP or wrap policy, the owner purchases that policy at the owner's expense. However, the policy is really for the benefit of the general contractor and the subcontractors, not just the owner. Should the general contractor and the subcontractors be obligated to contribute to the cost of the wrap policy? If so, how would the owner know whether the contribution from the general contractor or subcontractor is real, or whether the general contractor and subcontractors have simply increased their price to cover the additional cost? There is no clear answer to this question. There is also the question of deductibles. The wrap or OCIP policy will have a deductible or "Self Insured Retention" that the owner must pay. Should the general contractor or subcontractors be obligated to pay a proportionate share of the deductible since they are covered by the policy?

There are certain advantages and disadvantages to wrap policies. These policies are, at the time this book was written, relatively new on the market, and it is still too soon to determine whether they will function well in the marketplace. I suggest, however, that owners and contractors investigate the possibilities of wrap policies in place of the more traditional commercial general liability policies that are described in more detail a bit later on.

Liability Insurance

It is not uncommon for some damage or injury to occur during construction or for construction defects to occur or manifest themselves after the completion of the job. Examples during construction include a broken pipe that floods the project, an electrical short that causes a fire, a truck that backs into a wall, a stack of material that falls over damaging some other aspect of the job or injuring someone, and a visitor to the site who falls into a hole. All of these can and do happen on a construction site. Liability insurance is intended to protect the owner from this unexpected loss. In some circumstances, liability insurance will also offer some protection to the owner for construction defects.

To obtain this protection, several things must occur:

- The contractor or the owner must have a policy of liability insurance.
- The liability policy must have limits that are large enough to cover the damage.
- The premises of the project must be listed as insured property.
- The owner must be listed as an additional insured on the contractor's policy of liability insurance and on the policies of liability insurance issued to the subcontractors.

Accordingly, a well-written construction contract will contain specific language requiring the contractor to obtain and maintain liability insurance, with a specific policy limit, and it will require the contractor and subcontractors to name the owner as an additional insured under the policy. For additional protection to the owner, the contract should require the contractor and subcontractors to give the owner a copy of the entire policy, not just the additional insured endorsement (usually a one-page document that says that the owner is also insured under the contractor's policy). This practice is preferred because when the owner receives a copy of the entire contract it may be that years later if a claim is made, the contractor or subcontractor may have lost the policy. By having the entire policy, the owner can be assured that the policy has actually been issued, that the amount of coverage is correct, and that the property has been correctly identified. Having a copy of the entire policy makes it easier to identify the carrier and to make a claim, particularly when the subcontractor may no longer be in business or may be difficult to locate. Requiring the contractor and subcontractor to deliver the whole policy also enables the owner to verify that the promised coverage has actually been purchased.

In addition to its basic function of providing coverage for damage to the property and injuries suffered by people or things (such as vehicles) on the property, the liability policy can serve an additional important function. A liability policy with a *broad form endorsement* provides insurance coverage to the owner for construction defects caused by subcontractors when the work of that subcontractor damages the work of the contractor the work of other subcontractors, the property of other subcontractors or the property of the owner, and it also provides a mechanism for having an insurance company defend the contractor for claims of construction defects.

To provide at least some protection for an owner in the event that construction defects are discovered after the completion of the project, the policy must also have a *completed operations* coverage form. This means that the insurance coverage remains in effect for some period of time after the contractor's operations on the job are completed. In states with long statutes of limitation (California, for example, has a 10-year statute of limitations for latent construction defects), completed operations coverage with a 10-year "tail" (coverage will remain in effect for 10 years after completion of the project) would be preferable. For maximum protection, the "tail" should cover claims that arise "under the applicable law" of the state in which the coverage is issued. The difference between that language and using a 10-year yardstick to measure coverage is that some claims could be brought even after the 10-year statute of limitations, and the owner would want those claims to be covered.

Insurance coverage for construction defects is a topic that, again, has been discussed in volumes and volumes of materials and has been the subject of many legal decisions for a long time. Suffice it to say for purposes of this guide that if your contractor has a *comprehensive general liability policy* with a broad form endorsement and a completed operations endorsement, it will be much more likely that you will have an insurance company on the other end of a lawsuit for construction defects. This does not mean that you will automatically get paid by that insurance company, but it does mean that the contractor will have a lawyer appointed for him or her by the insurance company and that the insurance company will recognize that it will have to bear the large expense of legal proceedings in your construction defect lawsuit. This may help you recover at least some of your loss for construction defects.

Worker's Compensation Insurance

This type of insurance is somewhat easier to understand, and it is vitally important. Worker's compensation insurance provides coverage for any worker employed by you, the contractor, or the subcontractors, who is injured on the job. Without that coverage, if a worker employed by you is injured, not only might you have to bear the entire cost of the medical bills and lost earnings, it may be presumed that you were negligent and you may have a difficult time preventing that injured worker from collecting tens or perhaps hundreds of thousands of dollars from you. The well-written construction contract will contain a provision whereby each person who works on the job will have to provide you, the owner, with proof of worker's compensation insurance coverage.

Builder's All Risk Policies

This type of policy insures the contractor, and the owner if the owner is named as an additional insured under the policy, from a variety of property damage from events that could occur during the construction of the home. Here are a few examples:

- A subcontractor parks his truck on a slope above the property but doesn't set his emergency brake. The truck rolls into a stack of building materials and destroys the materials. The materials must be replaced and the progress of the job is slowed.

- A plumber is soldering a pipe next to a stud wall and starts a fire. A portion of the house burns down.

- The plumbing contractor is digging a trench for a water line and cuts through the roots of a large, valuable tree on the property. The tree dies and must be removed and replaced.

- A flood or other natural disaster occurs, and the construction to date is ruined.

Builder's all risk policies are a good means of protecting against these and related problems that occur during construction. Builder's all risk policies might not cover damage that occurs off the premises , such as damage to a tree that is caused by a piece of earth moving machinery being transported to the jobsite. That damage would be covered under the contractor's general liability or perhaps automobile policy. In projects that require a significant amount of large equipment, special attention should be paid to this consideration.

Anti-Construction Defect Provisions

Carefully thought-out language in construction agreements can prevent, or at least minimize, many of the most common construction defects. I believe that anti-construction defect language in contracts is severely under-utilized in the construction industry.

Here's how to deal with this. First, some of the language to avoid construction defects belongs in the contract between the owner and the general contractor. The general contractor has the overall responsibility for the proper and timely completion of the job, including the scheduling of the work and the supervision of the work of subcontractors to avoid faulty construction. Second, because the subcontractors actually do most of the work, separate contract language in the contracts between the general contractor and the subcontractors can be extremely effective in helping to prevent and minimize common construction defects. Third, whether the language is contained in the owner/contractor agreement or the contractor/subcontractor agreement, the contract provisions must be implemented if they are to be of any value.

Specific suggestions for language in the contracts between general contractors and owners and in the contracts with subcontractors are set forth following the discussion of common construction defects, to give you a context for the contract language.

Mechanic's Lien Contract Provisions

Everyone has probably heard the words *mechanic's lien* in connection with a construction project, although most homeowners do not clearly understand what a lien is or why it is important to have contract language that deals with liens.

A mechanic's lien is a remedy provided by the legislature to increase the ability of contractors and suppliers of materials to collect money for their work on or supplies delivered to a construction project. The lien is a document that is recorded with the county recorder in the county in which the home or building is located. The lien states the amount that the lien claimant (the contractor, subcontractor, or supplier) contends is owed. After the lien is recorded, the claimant has a certain amount of time within which to file a lawsuit (the time varies from state to state) to *foreclose* the lien, meaning that the property against which the lien has been recorded could be sold to satisfy the lien. It is this right to foreclose and force a sale of the owner's property that gives the lien holder leverage in a dispute over unpaid contractors' fees or charges for supplies.

Four kinds of people in the construction process can record liens: the general contractor, subcontractors, suppliers of materials, and architects and engineers. The requirements to file a lien are different for the general contractor than the subcontractors or material suppliers, and these differences are important to understand so that you are protected from surprises. The differences arise from the fact that the statutes creating mechanic's liens are also intended to put owners on notice that a subcontractor or supplier may be supplying services or materials for their project, even if they aren't dealing directly with that subcontractor or supplier.

Here is the way in which the owner is given notice. When a subcontractor or supplier provides labor or materials on a job, but that subcontractor or supplier does not have a direct contract with the owner, the subcontractor or supplier is required to give the owner a *preliminary notice*. The preliminary notice says that the subcontractor or supplier is supplying labor or material on the job. The owner now knows about the subcontractor or supplier. In order for the owner to be protected against a mechanic's lien from the subcontractor or supplier, the owner needs to obtain a lien release from the subcontractor or supplier.

In the real world, this is done by the general contractor, because the general contractor is paying the subcontractor or supplier directly. From the owner's perspective, he or she wants to avoid the following situation: The contractor requests $5000 to buy cement to pour a driveway. The owner, or construction lender if there is one, writes a $5000 check to the general contractor, with the expectation that the general contractor will pay the cement supplier. The general contractor has an account with the cement supplier, and orders $5000 worth of cement, which is delivered to the job site, but the general contractor doesn't actually pay the cement supplier the $5000. Now there is a dispute between the owner and the contractor. The contractor walks off the job. The cement supplier demands the money from the general contractor but can't collect. The cement supplier records a mechanic's lien against the owner. The general contractor cannot be found, and the owner might end up paying twice for the same cement.

To avoid this situation, the statutes dealing with mechanic's liens require the cement supplier to send the owner a preliminary notice. The owner then goes to the general contractor and requires that the general contractor obtain a lien release from the cement supplier in return for the $5000 check. If the cement supplier does not send the preliminary notice to the owner, the cement supplier cannot record a lien, and the cement supplier's sole remedy is to pursue the contractor for the payment.

Not all suppliers and subcontractors will send the owner a preliminary notice. It depends largely on their relationship with the general contractor. If the general contractor has a good payment history, the suppliers and subcontractors will rely on the general contractor's good faith. If not, or if the amount in question is substantial, a preliminary notice is likely to be sent.

The general contractor is not required to send the owner a preliminary notice (nor is anyone else who contracts directly with the owner) because the owner is already on notice that the general contractor (or anyone else with whom the owner contracts directly) is supplying labor and material for which the general contractor expects to be paid.

To facilitate this process this, a document called a *conditional lien release* is used. The conditional lien release is signed by the subcontractor or supplier and states that upon payment, the subcontractor or supplier releases any lien rights. When the check clears, the conditional lien release becomes unconditional. In situations where payments are made on account, the conditional lien releases for prior account payments become final when those payments are made.

With this background in mind, the contract terms dealing with lien releases become more understandable. Basically, the contract needs to say that the general contractor will provide the owner with lien releases for labor and material supplied by the general contractor as payments are made and with lien releases from subcontractors and suppliers for payments received by the general contractor for the benefit of those subcontractors and suppliers.

Usually, contracts with the general contractor require the general contractor to keep the project *lien free*. What this means is that the general contractor has to pay those people who supply labor and materials to the job and make sure that they do not record liens. If a lien is recorded, the general contractor is obligated under the contract to pay the amount owed and have the lien removed. Liens are removed by recording a document called a *lien release*. The contract should also provide that the contractor will pay the owner's legal fees to defend any foreclosure lawsuit filed by a subcontractor or supplier.

Indemnities

Construction contracts always have language dealing with indemnities. An *indemnity*, in the context of construction projects, is a promise by one person to hold another person harmless for injuries that may occur as a result of the work performed.

In a typical situation, the general contractor agrees to indemnify the owner against all claims of third parties that might arise from or be related to the work. Thus, if a person is injured on the job, and sues the owner, the general contractor will agree to defend and hold the owner harmless from the claim.

The issue that usually arises is the particular type of indemnity, and whether the owner has the right type of indemnity from the right person. This is of particular importance with regard to construction defects and later claims by third parties for construction defects against the owner (after a sale of the property by the owner).

An owner will want to have a particular kind of indemnity and will want to have it from two groups of people. The three types of indemnities are broadly called Type 1 (or Type A), Type 2 (or B) and Type 3 (C). A Type 1 indemnity says that the general contractor will indemnify the owner from claims even if the owner was the cause of those claims. Types 2 and 3 indemnities do not include that language, meaning that in any dispute between the owner and the contractor, the relative fault of the parties becomes an issue.

In the best of all possible worlds, the owner will always want to have a Type 1 indemnity from the contractor. The contractor, in turn, will always want to have Type 1 indemnities from all of the subcontractors.

Type 1 indemnities provide protection to the owner in two ways. First, if an accident or some other occurrence take place during construction that results in a claim, the owner will be protected by the indemnity. Second, if the owner sells the property and a later claim for construction defects is made, the owner will argue that the contractor and subcontractors are all obligated to defend and hold the owner harmless from those claims under the Type 1 indemnity agreement. If the general contractor has Type 1 indemnities from each of the subcontractors, the general contractor will be entitled to have the subcontractors defend and hold the general contractor harmless from the third party's claim as well. All of this will assist the owner in resolving the claim, because it will mean that the general contractor and all the subcontractors will have to participate in the defense of the claim and they will likely contribute to a settlement to resolve the claim.

3

Why Do Contractors Get Sued?

FOR OWNERS AND CONTRACTORS

Why do contractors get sued? The answers to this basic question provide the foundation for understanding how to avoid lawsuits, making them cheaper to litigate, and minimizing exposure.

Contractors get sued for the following reasons:

- They don't communicate with their customers and their customers don't communicate with them, leading to unfulfilled expectations.
- They don't finish their work on time and on budget, leading to frustration, anxiety, and anger.
- They don't finish their work, period.
- Their work is defective.
- Someone else gets sued and they get dragged into the battle.
- They pick the wrong clients.
- They agree to undertake work when they know that the owner doesn't have the money to pay.
- They take on work knowing that they aren't charging enough to get the work done and make a reasonable profit.
- They take on work that's outside their area of expertise or experience.

EVERY LAWSUIT STARTS THE SAME WAY

It sounds simplistic, but it's true: every single lawsuit begins with a dispute. The parties to the dispute cannot resolve that dispute by themselves. The dispute then ends up in court or arbitration, usually because the parties believe that they have no other way to resolve the dispute. (Later in the book, I discuss various methods of alternative dispute resolution to keep disputes from ending up in court or arbitration.) Experience shows that most disputes that arise from the construction process result from miscommunication. Effective communication is therefore one of the most important ways to avoid lawsuits from the construction process.

Common Areas of Poor Communication

Let's look at some of the typical areas of miscommunication and poor communication that lead to unfulfilled expectations, anger, frustration, and lawsuits. These are only a few examples, but they illustrate a common pattern in the communication between contractor and client.

Client: "You told me that everything was included in the contract and now I get a list of extras."

Client: "You told me the job would be finished by a certain date, and it's not done."

Client: "You told me that it wouldn't cost me more than $X.00, and now you say it's twice as much."

Client: "You told me that you paid all of the subcontractors, and now they tell me that they haven't been paid."

Contractor: "You told me that you had the money to pay for the job, and now you want me to wait to get paid."

Contractor: "You keep changing your mind."

Contractor: "You can't make up your mind."

Contractor: "Your designer/architect/interior decorator cannot provide plans and decisions for me."

The Nest Syndrome

To understand the communication process between a homeowner and a contractor, it is first necessary to have a context for the relationship. When a contractor is building or remodeling a single-family home, the relationship between the homeowner and the contractor isn't a strictly business relationship, because a home isn't a business: it's personal—very personal. When dealing with a home, an entire range of emotions comes into play for the homeowner, and those emotions have a powerful impact upon the relationship between the contractor and the client. I call it the *nest syndrome*. When a contractor and an owner don't understand that strong emotions are at work, the stage will have already been set for disputes. But by understanding and acknowledging this set of emotions, contractors and owners can understand and better deal with the many issues that arise during construction and avoid the disputes before they spiral out of control.

Probably the most common example of an emotional reaction that arises from the nest syndrome is the homeowner's desire to have the smallest and shortest disturbance of the nest. The effect of this is that delays and inconveniences that a contractor views as normal become disturbing and intolerable to the homeowner. Suppose, for example, that the owner expects that hardwood floors will be installed on a certain date. The owner comes home, only to discover that the hardwood flooring is stacked in the living room. The owner calls the contractor, upset. The owner might say: "Why aren't the floors down? You promised me that the floors would be installed by the end of this week."

The contractor knows that the floors aren't down because the supplier was late in delivering the floors, and the contractor wants the flooring material to season before it is installed. He also knows that the installation of the flooring is not on the critical path to completion of the house (meaning that a delay in the installation of the flooring will not cause a delay in the overall completion of the home). From the contractor's perspective, the delay is therefore of no importance.

But the contractor has failed to take into account the nest syndrome. To the owner, the delay in the floors is important for many reasons. First, to the owner, *everything* is on the critical path (of course, the owner doesn't know the term *critical path*, and doesn't use that term in his or her discussions with the contractor). The owner only knows that everything has to be done before the house is complete. To the owner, the flooring is a big item. If the flooring isn't being installed, it must mean a big delay or a big problem.

Second, remember that the owner wants his or her nest back as soon as possible. The floor is an important part of the nest. When the floor is installed, the project starts to look more like a nest and less like a mess. The faster the project begins to look like a nest, the happier the owner. (This is why many owners become unhappy and frustrated with the construction process during the time between the completion of rough framing and the completion of drywall—when the house is framed, the owner can visualize the nest. In fact, even though the owner knows, rationally, that it will take months to transform the rough framed structure into a house, at some emotional level, the owner sees a house that looks like it could be a nest fairly quickly, and the owner becomes disappointed when the transformation from framed-out nest to completed nest takes so long). Good communication on this specific subject (delay between framing and completion) can go a long way to having a more satisfied client, but it requires an appreciation for the client's emotional state and an understanding of the client's perspective on the construction process.

Remember that owners of single family homes likely have no prior experience with the construction process. Like anyone who is involved in something unfamiliar, owners are anxious because they don't know what to expect and they don't have the ability or background to rationally deal with the events. The general contractor, with years of experience, is accustomed to the process, with all of its delays, changes, uncertainties, setbacks, imperfections, disorder, and annoyances. The owner doesn't know that all of these are normal and may perceive each delay, change, setback, imperfection, and the like as a personal failing on the contractor's part, because the owner believes that the contractor has delayed the completion of the nest.

Finally, the other important part of the nest syndrome is that the owner is not living in the house while it's being built. The owner is living in another house, an apartment, with a relative, or even in a hotel. To an absolute certainty, that displaced owner wants out of the temporary arrangement as soon as possible. Therefore, delay not only means that the owner cannot inhabit the new nest, it also means that the owner has to continue living in an unfamiliar and uncomfortable temporary arrangement. Most people do not like temporary unfamiliar surroundings. The discomfort can make them short-tempered and irritable, and the contractor is likely to be on the receiving end of those emotions.

In the sections that follow, I will provide several methods of dealing with those emotions and keeping the relationship productive despite the strong emotions that owners typically experience.

An Action Plan for Better Communication: How and What to Communicate

Here is an action plan to improve communication.

Choose the Right Client

Every contractor gets excited when being considered for a new project. Every contractor looks to the potential profit in a new job and the possibility that it will lead to more work from the new client's friends and associates. No contractor likes to think about the potential problems. But, for contractors, choosing the right client is as important to avoiding a lawsuit as knowing the difference between a common nail and a box nail. As much as the client should assess the contractor and the contractor's qualifications, the contractor should do the same for the client. And if the client doesn't meet the standards, the contractor shouldn't take the job. Experience shows that it won't be worth it.

What should a contractor look for when evaluating a potential client? Here's a checklist:

- Are you the second, third, or fourth contractor on the job? If so, politely decline the job and move on to the next project.
- If the clients are a couple, do they have different needs, expectations, standards, and goals? If so, these differences need to be discussed and addressed at the outset of the job, not while the job is in progress.
- Is the client relying on an architect or designer that you think isn't competent? Even if your work is perfect, if it's based on poor designs or poor plans, the project will have defects and you will get the blame. Why start on a project where you will get blamed (and sued) for some else's bad work?
- Are you worried that the client doesn't have the money to pay for the project? Trust your instincts. If you are worried that the client doesn't have enough money, put the issue on the table. Don't be afraid to ask for proof of the client's ability to pay (you might want to ask to speak to the client's accountant, for example). If you work on a project and don't get paid, you would have been better off taking a vacation or looking for other projects.

- Does the client insist that you take the low bid for all of the subcontractors? Nobody can argue with the idea of trying to save money. A client's insistence on taking only low bids is a sign that the client doesn't care about quality, only about money, and that's fine—at least until problems with the job crop up. When problems occur, the same penny-pinching client might claim that you promised the best of everything and you could get sued for not providing it.

- Did your client choose you because you submitted the low bid? If your bid was reasonable (even if it was low), that's fine. But here you have to be painfully honest with yourself. If you underbid the job, that job is not going to end happily. The money that you "saved" in your low bid will probably be spent later.

Educate the Owners

The better the client understands the construction process, the better the relationship will be and the better the communication. Before you start the job, sit down with the owner and explain the process. Give the owner the benefit of your years of experience. Include the good, the bad, and the ugly.

Use the following checklist of issues as a guide to what you must discuss with the owner:

- *The permitting process, including changes that might be imposed by the governmental authority that could delay the start of the job, the completion of the job, or make the job more expensive.* Remind the owner that the permits are expensive, that there is a wide variance in the quality and attitude of plan checkers, and that plan checks might be affected by or delayed by increases in construction activity that are seasonal or based on an improved economic climate.

- *Unexpected site conditions that might delay the progress of the job, sometimes significantly, or increase the price (depending on the contract language).* Explain what site conditions are, because while you might know the term, it might be foreign to the owner.

- *Changes required by the inspector during the building process.* Explain that changes always delay the job, and changes can make the job more expensive. Also explain that inspectors are not given the job and do not undertake the job of looking for every possible construction defect on a project. Many owners, when they find construction defects after completion, ask why the inspector did not find them during the construction process. Owners become suspicious when told that the inspector missed it or when similar explanations are given, and they begin to suspect that there are other reasons (pay offs, etc.). Owners should know from the start that the inspector is looking only for certain things when he or she comes to the project, and owners should also be told that the city government cannot be sued for construction defects (this comes as a big surprise to most owners if there is a construction defect lawsuit after completion). In addition, owners need to know that simply having a permit does not mean that the job will be completed according to the plans. The building inspector has the last word on the construction, even if his or her opinion is different from the plans or the opinions of the architect or engineer. Decisions by the building inspector can cause delays, change orders, and increased expense (and often result from poor or missing plan details).

- *Potential delays.* Include delays in having the job inspected, and the fact that the job cannot progress until the inspections are completed. Also cover delays caused by late shipment of materials.

- *Problems with materials.* This includes substandard materials, wrong sizes, wrong colors, or damaged materials.

- *Acts of God.* This term might mean a lot to you, but probably means next to nothing to your owner. Explain that it refers to weather and other problems that are out of your control, such as strikes, transportation delays, or political upheaval. Specific examples are helpful. For example, in 2003, due to the war in Iraq, the price of plywood and steel rose dramatically, and these products became scarce. In 2004, because of construction of large dams and hydroelectric plants in China, cement, particularly white cement, became scarce and expensive in various parts of the United States.

- *Problems with neighbors.* Neighbors may complain about noise, dust, debris, boundary lines, height limits, fences, colors, and the like, and this can impede or even shut down a job.

- *Subcontractor scheduling.* Owners need to understand that a job progresses in a certain sequence. Once a subcontractor is delayed from completing its work, if that work is on the critical path, the job cannot progress in other areas. Remember that owners look at all the things that remain undone and are often confused why other work can't proceed—not understanding the need for sequencing and not understanding the idea of a critical path of the project.

- *Imperfection.* Owners need to understand that building materials aren't perfect. Lumber shrinks, warps, checks, splits, and twists. Buildings settle, cracks develop, adjustments may need to be made, colors don't always match, and thousands of small details cannot be made absolutely perfect. You know this well, but the owner may believe that he or she is getting an absolutely perfect home and may be unhappy if that expectation isn't met. The level of perfection is a critically important part of the communication process. You and the owner should have a clear understanding about the level of workmanship that is possible within the budget, and you should clearly advise the owner of where problems are likely to develop from shrinkage and settlement. Many owners find common shrinkage cracks and cracks from settlement to be alarming, and they interpret those cracks as evidence of serious problems with the building. By advising the owner in advance that they can and should expect these minor problems (and that they will be repaired as they appear), you have lowered the owner's level of anxiety and given the owner a more realistic set of expectations.

- *Dirt and debris.* Explain that construction is a messy business. Some owners don't care much about this and some are fanatically obsessed with a clean construction site. Know your client and respond accordingly.

- *The job will undoubtedly cost more than the original budget.* This just seems to happen, despite the best efforts of all concerned. An owner who is prepared in advance has less of a problem when the inevitable occurs. You should have a candid discussion with the owner and talk about how much he or she could afford if the project runs over budget. In fact, I believe that it is a better practice to advise the owner to have a reserve of at least 15 percent for overages, so that the owner can plan in advance for what seems to be the inevitable cost increases.

- *The "While we're at it, we might as well..." syndrome.* Nearly every client yields to the temptation to make changes, upgrades, improvements, and the like. But most clients seem to forget that all of these changes increase the price and increase the length of the work. It's your job to remind clients of the effect of these decisions, to keep the process on track.

Put It In Writing

Anyone who has ever played the child's game of telephone knows how imperfect verbal communications are. Words and phrases are misinterpreted, not heard, or forgotten. It is essential that all communications about important aspects of the construction process be documented in writing.

An example from a recent case illustrates the importance of writings. In this case, a general contractor was hired to build a 25,000 square foot custom home in an expensive area. The contractor specified the type of stone to be used to pave the outdoor areas of the home, and a subcontractor contracted to install the stone. Within a year after the installation, the entire stone installation began to fail. The stone faded, cracked, spalled, and pitted. The cost of repair was more than $500,000.

To avoid liability, the stone subcontractor took the position that prior to the installation he had informed the general contractor that the material supplied by the general contractor was not appropriate for the particular installation and had told the general contractor that the stone installation was certain to fail. Unfortunately, the subcontractor had no writing confirming this conversation and was, accordingly, exposed to considerable liability in the lawsuit.

This example shows the importance of documenting events that occur during construction. Arguably, even one memorandum, confirming letter, note, or email would have provided the subcontractor with an argument that the cost to repair the stone should be paid by the general contractor and not by the subcontractor.

Contractors are accustomed to written change orders, but many other matters need to be confirmed in writing, including the following:

- Substitution of materials
- Change of construction means and methods
- Decisions to reject the work of subcontractors or the goods delivered by suppliers (along with the delay that will occur from that rejection)
- Delay in the completion of the project, resulting from weather or other forces
- Problems with plans and specifications causing delay or other problems while the plans and specifications are clarified
- Instructions by the owner to the contractor with respect to particular aspects of the construction (for example, an owner says, "You have told me that this is not the best way to do this, but this is the way that I want it.")
- Decisions to change subcontractors, suppliers, or designers

- Price increases
- Daily reports that reflect weather conditions, identify the trades on the job, visitors, inspections, materials delivered, and all other material events that occur on a daily basis

The purpose of having all this information in writing is three-fold. First, it eliminates, or at least substantially reduces, disagreements later about the significance of changes in the work. Second, it helps the communication process by allowing the owner to have a clear understanding and expectation. Third, it eliminates the "he said, she said" argument years later in the process.

If you are a contractor, you are probably asking yourself how all of these items can be written down as the job progresses, without employing a full-time secretary. It isn't easy to keep track of paperwork, but here are some suggestions that will help.

For contractors, commercially available forms can be used. Building News supplies forms, and forms are also available from the American Institute of Architects. If you cannot find or don't like the commercially available forms, you can develop your own forms. Most contractors already use a form for change orders. These forms don't have to be fancy, but they do have to contain essential information. After you have developed the form, take it to a commercial printer and have the forms printed on three layer carbonless paper and bound at the top with adhesive (make them into a tablet with a cardboard backing). The top copy of the form should be kept in your job file. One copy should go the owner. The third copy should go the other party affected by the decision. Printing these types of forms is not expensive. (By comparison, think about this: most qualified construction lawyers charge at least $250.00 per hour. One hour of your lawyer's time will buy an awful lot of forms).

In addition to the information dealing with the specifics of the decision that you want to confirm in writing, make sure that the forms contain at least the following basic (but important) information:

- Job description (name of owner, property address)
- Date
- Signature lines for you, the owner, and any other party affected by the decision that is being confirmed in writing—i.e., subcontractor, supplier, architect, engineer, or other professional

Make sure that the form describes the subject matter of the discussion fully enough that someone reading it in the future would be able to understand it without an explanation from you or the owner. Be particularly careful to discuss the impact of the decision. A particular decision might mean that the job will be delayed or that it might be speeded up. It might mean that the final appearance of some portion of the job will be different from that originally contemplated. A decision might mean that the job will cost less or more money. A substitution of materials might mean no change in appearance or time, but perhaps the longevity of the material will be affected or increased maintenance will result. Thus, for example, the deletion of lightweight concrete floors in the second floor area of a residence would ordinarily speed up the job and reduce the cost, but it also might lead to more noise from floor to floor and less stiffness in the flooring surface. In that case, the language in your discussion about this change would include reference to any deduction from the contract price, any reduction in the planned length of the job, and special reference to noise problems, and perhaps reference to more limited flooring finishes on the second floor due to the more flexible floor assembly.

Finally, and this is discussed in more detail later, it is vitally important to keep the forms in your file and to keep your files organized and available even years later in the event of a lawsuit. Remember that in most states, the statute of limitations for latent defects from construction is long (California, for example, has a 10 year statute of limitations for latent defects). Your files should be kept by you for at least 10 years from the completion of the job because you could be exposed to a lawsuit for the entire 10 year period and your file contains vital information for your defense.

When Things Go Wrong

When things start to go wrong, contractors need to focus even more carefully on good communication. Unfortunately, some contractors stop communicating when things go wrong, just when they ought to be communicating well. Things go wrong on every job. Inspectors make demands that no one anticipated. Site conditions are different than expected. A key subcontractor closes up shop. A material that was just perfect goes out of production. Or somebody simply makes a mistake. And what happens? Delay, to a certainty, and usually additional expense.

Many contractors think that they can hide the problem. They believe that by scheduling a little more manpower on the job, they can make up the for the delay. They are also reluctant to confront the owner, because usually the confrontation involves some admission of responsibility, at the least, that a mistake was made. And nobody likes to admit mistakes.

But this is the best time to be up-front with owner. Owners make mistakes, too. In fact, most contractors would tell you that they believe that owners make lots of mistakes. Owners understand that mistakes are part of life. Most owners will not only respect you for having the courage to admit the mistake, but they will also increase their level of trust and confidence because they know that you are not hiding information from them.

Once you have discussed the mistake or other problem with the owner, if the mistake or problem will slow down the job or increase the price, take out your form book and start writing. Now you have not only addressed the problem, but you have documented the fact that the owner was advised and that everyone agreed on a plan to fix the problem.

Probably the best way to avoid these problems in the first place is to have weekly meetings, and to conduct those meetings according to a rather formal protocol (which includes notes). At a weekly meeting, the parties will discuss the progress of the job, any delays, any anticipated problems, cost increases, cost savings, labor issues, material issues, weather problems, milestones, and other events that materially affect the job.

The notes from the prior week should be reviewed. The meeting notes should contain an "action list" with specific tasks designated to specific individuals. Notes should be made whether those tasks have been accomplished on time, and if not, why.

This paper record not only assists everyone in keeping track of the many details that arise in any job, but also helps you, the contractor, protect yourself against claims by the owner during and after the project. Weekly meeting notes are *contemporaneous documents,* created at the time the events are occurring. If a dispute crops up later, the contemporaneous documents are always the most persuasive evidence for the person deciding that dispute, whether that person is a judge or an arbitrator (or a jury). It's part of what you might call "defensive contracting," because the first thing that the judge, arbitrator, or jury will want to know is whether you have something in writing. Your ability to say "yes" will go a long way toward reducing or even eliminating your potential liability in a lawsuit.

Effective Communication

Knowing what to communicate is different from knowing how to communicate. And knowing how to communicate is just as important in maintaining a good relationship between a client and a contractor. Let's take a step back from the specifics of good communication techniques and focus for a moment on a few other aspects of the nature of the relationship between a contractor and an owner, particularly in the construction of a single family home.

As discussed earlier in the context of the nest syndrome, the construction of a home, or even a commercial building, is a special business relationship. It is one of the few relationships where two relative strangers work closely together to create something that they expect to last for a long time, something that will be important economically, and something that will be important emotionally.

For these reasons, the relationship between contractor and owner is quite complex. Evidence of this is provided by the way in which owners often talk about contractors: "I love my contractor." Or "I hate my contractor." Rarely is there an in-between. Contractors tell stories, both positive and negative, about the demands made by owners, about the personal secrets, needs, desires, wants, problems, and difficulties of owners.

How do owners and contractors communicate effectively in this relationship? The answer lies in learning a new set of communication skills and using them during the construction process.

Let's start with a concept described often by professional mediators as *active listening*. Active listening is a technique whereby a conversation becomes an exchange of information, an opportunity for at least one of the parties to the conversation to express feelings and know that those feelings have been heard and understood by the other party, and to begin a process where disputes are avoided or resolved.

How does active listening work? First, in a normal conversation, the person who supposed to be listening is usually not listening intently to what the other person is saying.

Instead, the person listening is thinking about what response he or she wants to make to the points being raised by the person speaking. Here's an example:

The owner says, "This floor isn't what we ordered, it looks terrible, and I want it torn up at your expense." Hearing this, the contractor immediately begins to think how to defend himself. A likely response from the contractor might be, "You authorized that floor. We picked out the material with you. It was installed perfectly. If you don't like it, that's okay, but I'm not going to pay for it." The owner then reiterates that the floor is unacceptable. The contractor is puzzled by the owner's reaction, because he knows that the owner approved the floor, and he digs in his heels and insists, again, that he will not replace the floor at his expense. After a shouting match, the owner leaves the project, threatening to call his lawyer.

What would have happened if the contractor had been actively listening to the owner in this conversation? First, the contractor would have been listening not just for the words, but for the underlying concerns or emotions behind those words. Second, the contractor would respond to the underlying concerns and emotions. Third, the contractor would attempt to draw out the underlying concerns instead of asserting a defense to the points being raised by the owner.

Why would such an approach be effective? In most conversations in which the parties are raising concerns, the parties state *positions*. Positions, however, are different from the underlying *interest* that the party is trying to advance or protect. Go back to our example, and the owner's comment about being upset about the appearance of the floor. The owner's position is that the floor is unacceptable. But what is the owner's interest? The owner's interest might have little to do with the position that he has stated.

Suppose, for example, that the owner's wife insisted that the designer make all the decisions about the flooring. In the discussions with the designer, without the contractor being present, the husband says that he wants a particular floor. The designer disagrees, and warns the husband that the floor will look terrible. The husband insists, and reassures his wife that the floor will look great. The floor is installed and it looks terrible. Now the husband has a problem. He didn't take the designer's advice. His wife is angry. He needs to find a way out and decides to make the argument that the floor wasn't installed correctly, so that it can be replaced with a different floor at the contractor's expense. The owner's interest is saving face with the designer and his wife, and not losing the money that it will cost to replace the floor.

Now let's replay the conversation, this time using some active listening techniques.

Owner: "The floor looks terrible. It's unacceptable. This isn't what we ordered. I want it torn up at your expense."

Contractor: "You sound really upset about this floor."

Owner: "You bet I am. I can't live with this floor. Everybody hates it. The designer hates it. My wife hates it. Even I don't like it anymore."

Contractor: "It sounds like the designer and Mrs. Owner are giving you a hard time about this."

Owner: "A hard time? You won't believe what a hard time they are giving me. The designer knows I made a big mistake, and my wife is really upset because we spent all this money on a floor that doesn't look right."

Contractor: "This is not a good situation. You're upset, the designer's upset, and Mrs. Owner is upset. I think that anybody would be upset in this situation. I wonder what we can do to make this problem smaller and not bigger?"

Owner: "Look, I know that you used the material that we selected. But we are going to have to tear it up. What can we work out?"

Contractor: "Let me give it some thought. Maybe we can remove it at a reduced rate and substitute a less expensive replacement so that the whole thing doesn't cost so much. That way you can explain to the designer and Mrs. Owner that you were successful in getting the floor removed without your bearing the whole cost."

The parties are probably on their way to a resolution of this dispute, either on the basis outlined above or some other variation of it. But the important point is that by using the principles of active listening, the contractor was able to turn what could have been a major problem into a problem with a solution.

Whole books and many articles have been written about active listening and related effective communication skills, and a detailed explanation of the subject is much beyond the scope of this book. But I can provide you a basic approach to the subject. If you keep the approach in mind, and practice it, you will find that your communication with owners becomes more effective.

Here is an outline of active listening/effective communication techniques:

- Active listening requires patience, calmness, focus, and a lack of distractions. If you don't believe that you have all of those available to you at the time of the conversation, don't have the conversation. It's better to postpone the conversation than to have a poor conversation.
- To listen actively, search for key words in the other person's remarks, and repeat them back. Repeat the last two or three words the person said. If a person says, "I'm upset," repeat back, "You sound upset." By repeating words back, you are showing that you are listening and that you are hearing and understanding what the person is saying.
- Try to think less about what you want to say in response to what you are hearing, and think more about simply listening carefully to what the other person is saying.
- Don't try to solve the problem at the outset of the conversation. Many times, the person speaking doesn't want the problem solved; they want to talk about what they are thinking or feeling. Problem solving can come later in the conversation.
- Don't argue with the points being raised. Listen to the points. Acknowledge the points, by saying, "I hear what you are saying."
- Summarize your understanding. This is a good way to make sure that you have actually understood what the other person is saying, and a good way to tell the other person that you have been listening carefully.

- Search for the interest and avoid reacting to the position. A good way to do this is to say, "I can see you're upset. I need to understand better what is making you upset." Or, "I heard that you want me to fix this at my expense, but I want to understand why I should be the one to bear the cost."

- Commit to a resolution of the dispute. Say, "It seems clear that we have a problem. I want to resolve it, and I think that if we work together, we can resolve it."

- Steer the conversation away from confrontation when it becomes confrontational. One good way to do this is to ask, "How does that help us resolve our disagreement?"

- Build on the progress of the resolution. When the parties agree on at least part of the dispute, say so: "It seems like we agree on this part, what can we do to reach a resolution of the rest?"

- Be open to seeing both sides of the dispute.

DEALING WITH LATE AND OVER-BUDGET WORK

By now, you know one thing well: The longer it takes to complete a construction job, the more likely it is that the owner will become agitated, frustrated, and angry. Anger and frustration lead to disputes. Disputes lead to lawsuits.

Delay seems inevitable in construction, despite the best efforts of all concerned and despite incentive and penalty clauses. Since delay is inevitable, it is safer to address the problem from the start. Here is your action plan for avoiding disputes over delay.

Set Realistic Schedules   Contractors can be tempted to acquire work by giving owners an overly optimistic completion schedule. This should be avoided. It is better to disappoint the owner before the contract is signed by providing a realistic schedule than to disappoint the owner in the middle of construction by having to extend an unrealistic schedule.

Expect Inevitable Delays   Some contractors fear that telling the owner at the outset that the job will be delayed will make the contractor appear unprofessional or unable to manage the job effectively. The opposite is true. A contractor who clearly communicates to the owner that even the best effort of the contractor cannot overcome the delays that are part and parcel of construction will gain the respect of an owner for being honest and up-front.

Advocate Rapid Communication   This is part of the idea of not hiding out, as discussed earlier. When you first see a problem developing, that's the time to discuss it, not after the problem has already grown into a mushroom cloud over the project.

Basic Construction Law

All of the above can also be understood more completely with some context of the basic law of construction in mind. While this book is not intended to be a legal treatise on construction law (there are more than enough such works already in print), a basic overview of some of the more important issues is helpful.

Contractor's License Law Issues

Every state has some version of a contractor's license law because contracting is a heavily regulated business and is generally under the jurisdiction of some variety of consumer protection group in government.

These contractor's license laws are intended to accomplish several goals:

First, the basic process of licensing is intended to promote professionalism and competence for contractors. To this end, the various licensing laws provide that contractors are required to meet minimal standards of experience and expertise, demonstrated through written examinations among other things.

Second, the contractors license laws specify that contractors must operate their businesses and manage their relationships with customers in a manner that will protect the consumer of contracting services from incompetence and illegal behavior.

Third, the statutes are designed to provide the consumer with some recourse in the event that a contractor abandons a project, fails to pay subcontractors, performs shoddy work, or otherwise violates the important provisions of the various statutes. This recourse includes dispute resolution, usually in the form of a system under which the consumer may file a formal complaint against a contractor or have such a complaint heard within the administrative system of the contractor's board, and not in the courts.

Here are some important things to keep in mind with regard to contractors and the contractors law:

- The contractors licensing agencies are not perfect. Budgetary constraints make it difficult for them to act on many complaints, and their personnel are stretched thin, to the point that cases sometimes do not get the attention that they deserve.
- In some respects, although consumers may feel a strong sense of dissatisfaction with the agencies, contractors commonly feel that the deck is stacked against them and that all state agencies have a pro-consumer and anti-contractor bias. In addition, in many states, the mere fact of a complaint remains of record for many years, whether or not the complaint is upheld. Because complaints are more easily accessed over the Internet, the existence of a complaint, even if it is unfounded, can do serious damage to a contractor's reputation and business. For this reason, even the threat of a complaint in many instances may cause a contractor to respond more productively to an owner's demands for completion or repair of a project.
- A complaint to a licensing agency does not necessarily mean that an owner will prevail or that an owner will collect if the contractor is found liable. Many agencies, however, offer dispute resolution procedures that are less expensive and faster than court proceedings. In California, for example, the Contractor's State License Board provides for arbitration of disputes. If the owner elects arbitration of any dispute of $50,000 or less, the contractor is required to participate. The participation of a

contractor in arbitration of disputes of over $50,000 is voluntary. If the arbitrator makes an award in favor of the homeowner, the contractor is required to pay the award or suffer the loss of his or her license. However, because many contractors simply begin to do business under a different name, even this provision of the agency rules does not guarantee payment of an arbitration award.

- In many instances, if the agency forms the opinion that the dispute is essentially a civil matter better handled by the courts, the agency will do little, or nothing, to protect the interest of the homeowner.
- The filing of a complaint costs nothing, takes little time, and may result in a faster and more economical result in smaller disputes. For these reasons, it should always be considered as an option to resolve contractor-owner disputes.

Change Orders

Change orders, which are changes in the scope, cost, or time necessary to complete the work, seem to be an inevitable part of the construction process and, unfortunately, fertile ground for the growth of disputes and litigation between owners and contractors. The process of change orders, and how to avoid problems in the change order process, was discussed earlier in the chapter. For the purposes of an overview of the law of change orders, the following would be the most important categories of change order disputes:

- Disputes where the owner claims that the contractor is not entitled to request a change order. In these kinds of disputes, the owner typically claims that the change requested by the contractor is not a change at all; it is an item that was either explicitly or implicitly covered by the plans and specifications, and the contractor is not entitled to additional time or money for the item in question.
- Disputes where the contractor is entitled to ask for a change order, but the amount of time or money is perceived by the owner to be excessive.
- Disputes where the change order is not in writing, and the parties disagree over whether it was authorized or how much the contractor is entitled to be paid or provided additional time. Despite the fact that a good construction contract will have explicit language to the effect that no change order is effective unless it is in writing, and signed by the owner or the owner's agent, the problem of verbal change orders plagues the construction industry.
- Disputes where so many changes have occurred that the contractor claims that the fundamental nature of the contract has been changed. This is called a *cardinal change order*. In these types of disputes, the contractor typically attempts to renegotiate the entire contract or claims that the owner has breached the agreement by transforming the nature of the project.

In instances where the dispute centers on whether the contractor is entitled to request a change order, the legal principle is simple in concept but sometimes difficult to apply as a practical matter. The legal principle is that a contractor is not entitled to an *additive* change order (which results in more money or time to the contractor) if the item in question was within the scope of the original contract. The practical problem is nearly always a disagreement about what was included in the original contract.

For example, the framing subcontractor completes the rough framing on a new home. The framing subcontractor leaves a mountain of small pieces of framing lumber in the front yard. The owner wants the scrap lumber removed. The framing subcontractor says that disposal of the scrap lumber is not part of his contract. The contractor takes the same position. A review of the contracts discloses that neither the contract between the owner and the general contractor nor the subcontract between the general contractor and the framing subcontractor specifically discuss the removal of the scrap lumber.

The owner tells the general contractor that it was always the owner's belief that the scrap lumber would be hauled away as part of the framing operations. The owner also points to general language in the contract stating that the job will be kept clean and in good order. The contractor protests. The subcontractor protests. The parties are in a dispute.

Disputes over the amount of time and money to which the contractor is owed are governed by a different legal principle. In these kinds of disputes, the law provides that the parties should first look to the contract. A well-written contract will provide a protocol for the assessment of time and money change orders. For example, the contract might state that the contractor is entitled to charge on a time and material basis for all change orders, and it may further specify the labor rates and the markup, if any, on the cost of materials.

The problem usually arises when the contract is not entirely clear on how change orders should be priced. In these instances, the parties may be relegated to arguments over "reasonableness," a concept that in theory works well but in practice is rather ill-defined.

Disputes arising from unauthorized change orders can present major problems on a job. It is often the case that the contract between the parties will contain strict provisions that require the written approval of the owner or the owner's representative before a change order is effective. It is equally the case that in a job that is progressing rapidly, the formalities of the contract may be disregarded by the parties so as to avoid impeding the progress of the job. Usually the contractor says something like, "The architect approved the change on the job site and said that he would send the change order over later, but it never got done." Usually the owner says something like, "I never approved that change order, and I never would have approved it, because it was included in the scope of the original contract."

There are two conflicting legal principles that apply to these disputes. The first is that the contract governs the relationship of the parties. If the contract has provisions that require change orders to be in writing, in many instances that should, and will, end the discussion. However, many times there are equitable reasons why it would be unfair to allow the owner to obtain the benefit of the changed work without paying for it. In these instances, the equitable concept of *quantum meruit*, or the *reasonable value of the services* of the contractor, will dictate a different result.

When a contractor has supplied labor and material for a change order that wasn't signed, the contractor nearly always argues that the owner obtained value from that labor and materials. Equity perceives that it is unfair for an owner to keep the benefit and not pay for it.

The tension arises from the fact that the owner will immediately point to the terms of the contract that strictly prohibit any changes not approved in writing. In addition, the owner will argue that there was a reason for the contract language (to prevent the very sort of dispute in which the parties now find themselves). This can be a persuasive argument, especially when the change orders are submitted well after the fact and it's too late to determine what the real cost was. The trier of fact, whether judge, jury, or arbitrator, is then called upon to determine whether it is fairer to hold the contractor to the strict language of the contract or to force the owner to pay something for the benefit that the owner received.

One way to avoid such disputes is to include contract language that prohibits the trier of fact from considering any evidence of *quantum meruit* claims and other equitable arguments. If an owner can obtain a contractor's consent to this type of contract provision, these kinds of conflicts can be minimized, if not avoided entirely.

Cardinal change order disputes are somewhat more unusual, but they do occur. In these kinds of disputes, the contractor claims that the entire contract has been abrogated, because the job that the contract bid is not the job that the contractor has been asked to complete. When this occurs, the parties have a major dispute. Cardinal change order disputes can arise from only one change order (although this is unusual, and the change order would have to affect dramatically the scope, cost, and time of the project), but they usually arise from numerous change orders.

Prompt Pay Statutes

In most states, contractors obtain some statutory protections from situations in which owners refuse to pay through *prompt payment statutes*. These statutes provide that when there is no material disagreement with regard to how much is owed to a contractor, an owner who fails to pay undisputed amounts in a prompt manner (defined by the statute) will be liable to pay a substantial interest charge on the unpaid amounts.

The dispute over prompt payment usually centers on whether the amount in controversy was, in fact, undisputed. Naturally, the contractor takes the position that there was no disagreement or dispute. The owner takes the position that the amount was indeed disputed. The statute creates exposure for owners in the sense that if the contractor makes a prompt payment statute claim as part of a larger claim for unpaid monies, and the contractor prevails on his or her arguments, a substantial interest charge can be added to the unpaid amounts owing to the contractor.

Warranties

The law recognizes two basic types of warranties in construction: express warranties, and warranties that are implied in law.

Express warranties are written warranties provided by the general contractor, the subcontractors, or suppliers of materials. These types of warranties generally provide that the person supplying the warranty will repair or replace the defective item or work for a certain period of time. An express warranty could provide that the defective item or work would be repaired at no cost for the labor or materials, or that certain labor rates would be charged.

Express warranties given by general contractors and subcontractors are subject to negotiation. Although the custom and practice in many areas is to have a one-year warranty, it is not uncommon for two- or even three-year (or longer) warranties to be provided. In contrast, warranties from manufacturers of equipment (such as appliances or hvac equipment) are generally not negotiable.

Implied warranties are a different matter. The law implies that every product will be reasonably suitable for the purpose for which it was intended. In construction, the rule is no different. A building, whether it is a house or a commercial structure, is intended to be fit for habitation. The importance of implied warranties is that they enable the owner of a defectively built structure to sue for breach of implied warranty after the expiration of the express warranty in the contract. Breach of implied warranty is therefore another, alternative, means of collecting damages for defective construction.

Breach of Contract

Most construction contracts, as noted earlier, contain the essential terms governing the project: scope, time, payment, cost, and quality. The failure of either party to the contract to adhere to its terms can result in a lawsuit claiming breach of the construction contract.

Generally speaking, when a contractor sues an owner for breach of contract, the contractor claims that the owner has refused to pay. Sometimes the contractor claims that the owner has failed to cooperate in other ways, such as failing to make the property available or failing to provide assistance from the owner's representatives, including the architect.

When owners sue contractors, the claims usually arise from the contractors failure to get the job done on time, for failure to pay subcontractors, or for defective workmanship.

The legal principles with respect to breach of contract actions are not particularly complex. The trier of fact, whether judge, jury, or arbitrator, is required first to determine the rights of the parties from the contract itself. A well-written contract should have terms that are clear and that cover all of the possible areas where disputes may arise.

However, even a well-written contract cannot by itself resolve all the issues. An example will illustrate this. Suppose that a couple hires a contractor to build a new custom home. There is a large art studio space in the new home because one of the parties is an artist. The couple informs the contractor that the artist will be renting a studio while the home is being built, and that it is important that the home be finished by a date certain so as to minimize the expense of renting this commercial studio space.

Of course the job takes longer than anyone expects. At the end of the job, the contractor submits his final bill. The owners want to deduct six months of rent for the commercial space. In one sense, the resolution of the dispute should be simple: the contract provides that the job was supposed to be completed by a certain date and the law provides that damages for additional costs are recoverable by the owners. But the contractor claims that the owners failed to make decisions about finish items that had a "long lead time." A long lead time item takes a good deal of time to arrive on site after it has been ordered (a special order cabinet, for example). The contractor argues that he is not responsible for the delay and refuses to pay the owners for the rent on the studio.

Oftentimes, other, more subtle, factors affect the outcome of this kind of dispute. Among them are the following:

- The contractor may need to get the last payment to pay for subcontractors or suppliers, or the last payment may represent the contractor's entire profit in the job. For this reason, the contractor may compromise and take less in order to get paid something.

- There may be an attorney's fees provision in the contract, meaning that if a lawsuit ensues, the *prevailing party* in the lawsuit (the party who wins) would be entitled to attorney's fees. The attorney's fees provision creates a good deal of leverage in any dispute, because the losing party ends up paying two sets of legal fees: their own legal fees and those of the opposing party.

- The contractor might want to bargain over some other aspect of the matter and may be willing to compromise on the issue of fees. For example, perhaps the contractor provided a two-year warranty but wants to reduce the warranty to one year. The contractor might be willing to accept less in return for a reduction of the warranty period.

Negligence

The law provides that an owner, or in some instances a subsequent owner, of a property will have rights against a contractor for negligence in the design or construction of a building or home.

When an owner contracts directly with a contractor, the owner will have a variety of legal remedies against the contractor, including breach of contract, strict liability (if the contractor or developer is a "mass producer" of homes), negligence, and breach of warranty. To prove negligence, the owner must demonstrate, usually through expert testimony, that the contractor's construction of the building was either below the standard of care for similar buildings in the same area or that the construction violated a building code, manufacturers specification, or the provisions of the contract.

In some instances, "downstream" purchasers of a property—i.e., people who did not contract directly with the owner or developer—can sue the contractor for negligence. In these instances, the subsequent owner must satisfy a multi-part legal standard to determine the severity of the harm, whether the owner was within a class of individuals to whom the law affords protection and whether the contractor could have anticipated that his failure to act properly would harm subsequent owners.

Background to Construction Defect Law

FOR OWNERS AND CONTRACTORS

In the 1980s, developers realized that a fortune could be made by building structures that were basically similar to apartments, but selling them as condominiums. Thousands of condominium projects sprang up, mostly on both coasts of the United States and adjacent to large urban areas. Shortly after the condominium construction boom occurred, the construction defect litigation boom followed.

Many commentators have offered opinions on why construction defect litigation blossomed. I believe that there are many reasons, among them the following:

- Developers built condominium projects in the same way that they built apartments, but they failed to realize that the mentality of an owner of a condominium is far different from the mentality of a tenant of an apartment. Tenants will tolerate defects and simply call the landlord. Owners have a much lower level of tolerance and will act to protect the value of their investment by suing builders and developers.

- Because many condominium projects are quite large, involving hundreds of units, the owners could band together to bring lawsuits that would have been prohibitively expensive for only one owner.

- The developers were so consumed with the pressure to build new projects, and to build them at the lowest possible cost, that they employed unskilled labor and failed to hire trained and experienced supervisors. Builders failed to pay particular attention to two issues, water intrusion and noise, both of which were alarming to owners and both of which triggered most construction defect claims.

- Multi-family "production" housing is difficult to build without defects unless careful attention is paid to details and careful job supervision is provided.

- A small number of sophisticated lawyers recognized that changes in the law, primarily a change in the law that makes builders of multi-family housing strictly liable for construction defects, made it relatively easy to bring and win construction defect lawsuits.

- The "word" about construction defect lawsuits spread rapidly in the legal community and in the community of condominium owners and managers. Condominium associations and managers became acutely conscious of the need to look for construction defects and to file lawsuits to avoid statutes of limitations barring such claims.
- The 10-year statute of limitations in California for latent construction defect lawsuits is a long statute, allowing ample opportunity for defects to manifest themselves.

WHAT IS A CONSTRUCTION DEFECT?

What is a construction defect and why are construction defects still so prevalent in the construction industry? A construction defect is a defect or deficiency in the design, construction, or materials on a construction project. Broadly speaking, construction defects fall into two categories: defects that affect the performance of the structure, and defects that affect the appearance of the structure. Defects that affect the performance of the structure can be categorized, broadly, in two additional ways: defects that allow water into the structure, and defects that make the building structurally unsound or weak, and thus less resistant to wind and earthquake. Although there are dozens of examples, a few will suffice:

- A window or door that was installed without proper weatherproofing so that water intrudes into the building would be a construction defect.
- A balcony that is sloped toward a house so that water drains into the house when it rains would be a construction defect.
- A shear wall that was not nailed properly would be a construction defect.
- A foundation slab that was installed without a vapor barrier underneath to prevent water from coming up through the slab would be a construction defect.
- Soil that is compacted less than the degree of compaction required by the soils engineer would be a construction defect.

From the legal perspective, a construction defect is defined in somewhat different terms. Legally, a construction defect is a violation of the applicable building code, a violation of the standard of care in the community in which the project is located, or a violation of the manufacturer's recommendations. The *standard of care* means the way in which contractors in the same community, for the same basic sort of structure, install the particular component, although the standard of care will not permit a contractor to violate the building code, even if everyone in the community violates the building code with respect to the issue in question. In California, construction defects are defined by statute.

THE LAW OF CONSTRUCTION DEFECTS

In the 1960s, a developer named Eichler built a number of suburban homes in Palo Alto, California. The Eichler homes were quite unusual in their design, and one of their interesting features was radiant heat in the concrete slab floors. This heat was produced using hot-water filled tubes embedded in the slabs. Because of material shortages caused by the Korean war, galvanized steel pipe was used instead of copper to carry the water under the slabs, and the construction techniques used were of poor quality. As a result, the systems began to fail on a wholesale basis. One of the owners sued Eichler for construction defects. The case made its way to the California Court of Appeal, and the Court issued a now-famous decision on the law of construction defects (Kreigler v. Eichler Homes (1969) 269 Cal.App.2d 224, 74 California Reporter 749). The Court found that Eichler was a "mass producer" of housing (because scores of Eichler homes were built). The Court reasoned that there should be no difference between a person or company that is a mass producer of houses and a person or company that is a mass producer of any other sort of product. From this, the Court found that the law of products liability should apply to defective construction.

This decision was important because in California, as in most other states, since the 1960s, the law has said that mass producers of products are "strictly liable" for their defective products. When a theory of strict liability applies, the injured person only needs to show that the product is defective and that there has been damage or injury. The person does not need to show negligence, nor prove breach of contract or breach of warranty.

Prior to the Eichler case, the law of strict product liability had not been applied to housing in California. In other jurisdictions, courts had employed a similar analysis in the context of personal injury claims arising from defective construction (for example, Schipper v. Levitt & Sons (1965) 44 N.J. 70, 207 A.2d 314). Before Eichler, a person claiming construction defects would have to show either negligence, breach of contract, or breach of warranty to recover. Now, a person claiming a construction defect need only prove that such a defect exists. The person claiming the construction defect typically does so by demonstrating that the installation violates the building code, the standard of care, or the manufacturer's recommendations and specifications for the particular building component. In California, for properties constructed after 2003, the standard of proof for claims of construction defects in residential construction is even less. Under a comprehensive construction defect statute enacted in 2002, California law provides a broad and inclusive list of construction standards (California Civil Code section 896 [Standards for Residential Construction]). To prevail in a construction defect action, a plaintiff is required only to show that the actual construction does not meet the standards (California Civil Code section 942).

The damages that are recoverable in a construction defect case, at least in California (although most other jurisdictions are similar), are the lesser of the reasonable cost of repair or the diminution in value caused by the defect (Heninger v. Dunn (1980) 101 Cal.App.3d 858; Orndorff v. Christiana Community Builders (1990) 217 Cal.App.3d 683). At least one California case stands for the proposition that the "lesser of" rule does not apply to construction defects, but this decision appears anomalous (Raven's Cove Townhomes, Inc. v. Knuppe Development (1981) 114 Cal.App.3d 783). Under the new California construction defect statute, the damages for defects in residential construction are controlled by statute and basically consist of the cost to repair, plus additional damages for the cost of investigation and the cost to relocate while repairs are made (California Civil Code section 944).

Because builders of mass-produced housing (under California law, usually building three or more homes in a five year period makes a builder a mass producer of housing) are strictly liable, and because the level of proof under strict liability is so low, the battleground in most construction defect cases is the method, scope, and cost of repair.

In every construction defect case, the plaintiff claims a number of defects. The plaintiff retains experts who testify about the nature and extent of the defects. These experts also testify about the method, scope, and cost of repair. The defense—the developer, general contractor, or subcontractors—also have experts. The defense experts evaluate the plaintiff's claims and they examine the building to determine whether the defects exist; if they do, they determine the extent of the defects and an appropriate method, scope, and cost of repair.

What does *method*, *scope*, and *cost of repair* mean? Consider an example. Let's say that the stucco on a home is badly cracked. The owner retains a stucco expert (yes, there are stucco experts). The expert examines the stucco system and may remove pieces to determine how the stucco was attached to the building, and decides that the stucco subcontractor put too much sand in the mix when the stucco was applied. The owner's stucco expert then states that the method of repair is to remove all the stucco, the scope of repair is to remove it from every place on the house, and the cost of repair is $X.00.

The defense expert examines the building. He discovers that while there is, in fact, too much sand in the mix, the stucco was applied on three separate days. The mix was improper only on the first day, as the second and third days' applications used a different mix. He also believes that even though the mix was bad the first day, the stucco does not need to be removed. He states that on the portion of the house with the bad mix, the method of repair should be to sandblast the finish and recoat it, and that no fix is required on the portion of the house that was stuccoed on days two and three. He might also disagree on the cost proposed by the owner's expert.

WHY DO CONSTRUCTION DEFECTS OCCUR?

Here's an old joke:

Q: What's the difference between ignorance and apathy?

A: I don't know and I don't care.

Much the same can be said of construction defects. The cause of virtually all construction defects can be attributed to builders who either don't know (i.e., they act out of ignorance) or builders who know, but don't care (i.e., they act out of apathy). In my experience, it is almost always a little of both. Many builders never learn good construction techniques or basic construction principles (one example is the seemingly large number of builders who apparently believe that stucco is waterproof). Some builders learned good techniques, but they haven't kept current with new construction technologies or new understanding of better construction methods. (One example would be the use of exterior insulated finish systems, a one-coat synthetic stucco system that was originally thought to provide protection from water intrusion when applied in a certain manner but is now understood to fail unless much different application techniques are used.)

Apathy is also common. Whether the builder is pressed for time, or cost constraints are making it difficult to take the time, hire the right people, and use the right materials and techniques, many builders seem to simply give up. Some people call it the "good enough" syndrome: the builder knows that the job wasn't really done the right way, but it was "good enough." Good enough rarely is, and almost always results in a failure of some building system and a construction defect lawsuit.

The results of the attitude of "I don't know and I don't care" can be quite remarkable. Some photographs will illustrate. Figure 4-1 shows a pipe penetration at a roof. The builder carefully installed a sleeve around the pipe, but failed to waterproof the gap between the sleeve and the pipe, leaving a clear path for water to intrude into the structure.

Figure 4-2 shows another pipe penetration at a roof. In this photograph, the builder has carefully wrapped the foam insulation around the refrigerant lines of the hvac system to protect the foam from UV rays, but he or she stuck the foam wrapped lines through the roof. As soon as water saturates the foam insulation (which will take only moments), the water will travel through the foam into the structure.

A variation of the same approach to carefully sealing a roof penetration, but using entirely the wrong material, is shown in Figure 4-3, where a builder "sealed" around an hvac line by using expanding foam (a porous material).

Figure 4-1

Improperly waterproofed pipe allows water intrusion

Figure 4-2

The foam insulation will allow water to intrude

In Figure 4-4, the builder installed a screed at the bottom of a stucco wall. Unfortunately, the builder did not seem to understand the purpose of the screed (which is to allow water to drip or "weep" out of the stucco). Instead of using a weep screed, the builder used a *J* mold, which will effectively trap all the water at the bottom of the stucco wall and eventually cause water to leak into the structure.

On the subject of builders not understanding that stucco is porous, Figure 4-5 shows an effort to waterproof below a window by applying caulk and a metal flashing material

Figure 4-3

Builder "sealed" around an hvac line using expanding foam, which is porous

Figure 4-4
Improper J mold will cause water to pool and leak into the structure

Figure 4-5
Applying caulk and flashing on stucco is useless

Figure 4-5
(*Continued*)

on top of the stucco. The builder apparently failed to realize that since stucco is porous, applying caulk and flashing on top of the stucco is completely useless.

The results of poor quality construction can be dramatic. The four photographs shown in Figures 4-6 through 4-9 show a single family home that suffered severe dry rot from poor flashing details around windows. The dry rot in this residence was so severe that

Figure 4-6
Water intrusion at the exterior of this residence has resulted in severe dry rot to the exterior window bay joists and garage header

Figure 4-7
Water intrusion has damaged the framing around the bottom of the exterior windows to the point where the framing members are almost completely rotted away

Figure 4-8
The plywood shear panels are completely rotted through from water intrusion

Figure 4-9

In this photograph the 8x14 garage header has been almost completely destroyed, along with the framing and plywood shear panels at the lower edge of the windows. This home is no longer structurally sound.

when the stucco was removed from the exterior, no shear walls were found. It was first assumed that the builder had negligently omitted the shear walls, but further investigation revealed that the shear walls had completely rotted away (in a house that was less than five years old), leaving only the framing underneath.

5 The Most Common Construction Defects and How to Avoid Them

FOR CONTRACTORS AND OWNERS

At the end of the movie *Casablanca*, the French police officer character played by Claude Rains turns to his deputy and says, "Round up the usual suspects." It is much the same in construction defect litigation. At least half, and probably three-quarters, of all disputes between owners and contractors in residential construction involve the same construction defects, and they occur over and over again. They are the "usual suspects" of the construction world. That's the bad news. The good news is that because most construction defects fall into typical areas, it is easy to avoid creating construction defects in the first place.

HOW CONSTRUCTION DEFECT CLAIMS START

Most non-structural construction defect cases start with water intrusion. For owners, water intrusion is not only annoying, it is alarming. When the problem gets worse, or it doesn't get fixed, owners usually contact an expert or an attorney, and the next step is a lawsuit.

After the lawsuit starts, a good construction defect lawyer will hire experts who will carefully investigate the condition of the structure to discover every possible defect. They will then develop a scope, method, and cost of repair. The cost of repair often exceeds the original cost of construction. Worse, if the water intrusion has gone untreated for any period of time, mold may have developed. Mold claims greatly increase the cost of repair and they can also create a separate claim for personal injuries. Claims and damage awards for mold-related personal injuries can be very substantial, running into the millions of dollars (for example, in November 2005 in Los Angeles, a construction defect/mold personal injury in which the plaintiff alleged that a newborn's autism was caused by mold settled for $23 million).

The best way to avoid construction defects is to understand which defects commonly occur. The actual construction practices to avoid the defects are, in many ways, the simplest part of the process.

THE MOST COMMON CONSTRUCTION DEFECTS

What follows is a list of the "usual suspects" that lead to construction defect lawsuits and suggestions for ways to avoid making the mistakes that cause the defects in the first place. This section is organized (literally) from the ground up, starting with soils issues and ending with roofing issues.

For each category of defect, background information is provided pertaining to the construction issue, followed by a description of the typical defect and an action guide to avoid the defect. You'll find "anti-construction defect" forms for use in the owner/ contractor agreement and the contractor/subcontractor agreements. To a certain extent, these anti-construction defect forms summarize the "usual suspects" of construction defects.

DEFECTS IN SOIL AND FOUNDATION

Defects in Preparation of Underlying Soils

Soils problems tend to fall into three main categories: soils that are saturated with water, soils that are expansive, and soils that are not properly compacted.

Saturated Soils

In many locations, the water table (i.e., the level of naturally occurring ground water below grade, or the top surface of the soil) may be surprisingly high. High water tables can result from naturally occurring ground water, springs, or drainage from slopes. When the water table is high, special provisions must be made to protect the building from water intrusion at the foundation level, and special provisions must be made to insure that the foundations are stable.

These provisions can include the installation of drainage systems to prevent water from entering the soil, drainage systems to remove the water from the soil and transport it away from the building site, or drainage systems that collect water into pits, or sumps, from which it is pumped away from the building site.

An additional problem with saturated soils is that excessively wet soils cannot be compacted to the point that they are firm and stable enough to support a conventional foundation system. Efforts to compact excessively wet soils simply result in the soil being moved around underneath the compacting machinery, but not becoming compressed, a process called *pumping*. When this occurs, three options are available: remove the source of water and let the soil dry out sufficiently so that it can be compacted, remove the saturated soil and replace it with dry soil (and remove the source of water), or design and install a foundation system that does not rest on the saturated soil.

Expansive Soils

Expansive soils contain a high percentage of clay. Clay has a crystalline structure. When water is introduced into clayey soil, the water penetrates the spaces between the crystals and the crystals move apart. This, in turn, causes the soil to expand. The force of expansive soils can be enormous. It is quite common to see entire foundation systems broken and heaved due to expansive soils. Where expansive soils are present, special consideration must be given to the foundation system so that the foundations will withstand the force of the expansive soil.

The options are somewhat limited: the expansive soil can be removed and replaced with non-expansive soil, an engineered foundation system can be designed and installed to withstand the upheaval forces of expansive soil, or when expansive soils are left in place and used with engineered foundation systems, water management systems are used to prevent the introduction of water into the expansive soils.

Poorly Compacted Soils

Improperly compacted soil presents exactly the opposite problem. To provide adequate support for a foundation, whether a slab on grade foundation or a raised foundation, the soil under the foundation must be sufficiently compacted so that it supports the weight of the structure above. If the soil cannot support the weight of the structure, the soil will begin to subside, and the structure that sits on the soil will subside along with the soil under it.

Generally speaking, soil needs to be well-compacted to support the weight of a structure. What does *well-compacted* mean? To answer that question, you should first understand something about what soil is and how it is compacted. Soil is a combination of small particles of sand, clay, and rock. In most soil, some organic matter is also present, such as roots and small particles of plant material. In all soil, spaces between the small particles of sand, clay, and rock, called *interstices*, are present. When soil is compacted, a mechanical outside force is pressed onto the soil, forcing the small particles more closely together and decreasing the size of the interstices. Water is introduced to the soil during this process to act as a lubricant, because the small particles of sand, clay, and rock are irregular and the lubricant helps the particles slide together.

The maximum compaction for a given type of soil is determined through a laboratory test, in which the optimum moisture content is introduced into the soil (e.g., the precise amount of water needed to lubricate the particles of soil), and a mechanical means of compaction (usually a device that pounds soil into a tube) is used to compact the soil until no more compaction can be achieved. A measured amount of the thoroughly compacted soil is then weighed. The weight of the highly compacted soil is expressed as *maximum theoretical dry density*.

To determine what level of compaction is necessary for the soil under a building, the soils engineer takes samples of the soil from different areas at the site. She determines the maximum theoretical dry density for each different type of soil that she finds at the site. She then instructs the grading contractor that the soil at the site is to be compacted to at least 90 percent of the maximum theoretical dry density. To verify whether this has occurred in the field, the soils engineer takes samples of the compacted soils and then weighs a given amount of the soil. The sample must weigh at least 90 percent of the theoretical dry density for the same type and amount of soil.

The compacting process in the field is achieved using mechanical devices to press down on the soil. Large metal rollers can be dragged back and forth across the soil, or machines can be used to pound or tamp the soil down. Regardless of what device is used, soil is compacted in layers, called *lifts*, to assure that the mechanical device has compacted all of the soil in a particular layer. To confirm that this has occurred, the soils engineer will test the compaction after each lift, which may be as thin as 6 inches or as thick as 18 inches, depending on soils conditions. A record, or log, is kept of these tests to verify that the soil has been properly compacted.

Many things can adversely affect the compaction of soil. If too much organic material remains in the soil, it may compact, but it will not perform well over time, because as the organic material deteriorates, it leaves voids, or pockets, in the soil, and the soil above the void will tend to collapse into the void. Large rocks will interfere with the ability of the soil to perform well. A lack of moisture during the compaction process will cause less then optimal compaction because, as noted, water is necessary to act as a lubricant to force the particles of soil together. Too much water will also prevent proper compaction.

Because the benchmark to determine 90 percent relative compaction is the theoretical maximum dry density of the soil, and because the maximum dry density varies from soil type to soil type, if the soils engineer does not obtain enough samples from the site, her benchmark figures may be incorrect. If this occurs, the soils engineer and grader may be assuming that they have successfully compacted the soil when in fact the soil is poorly compacted.

Finally, many things happen in the field. A contractor who is anxious to get the job done might install a thicker lift than called out by the soils engineer, or he might install two or even three lifts without testing, and either mislead the engineer or persuade the engineer that the lifts do not require testing. If those lifts were not properly compacted, future trouble may be imminent.

Related to this is the issue of licensed grading contractors. Many contractors who lack a grading contractor's license, to save time and money, will use their own personnel and equipment to grade and compact soil. The Business and Professions code in California, and in many other states, makes this practice unlawful. This has a significant impact in a lawsuit, because the plaintiff in a construction defect lawsuit is entitled to a jury instruction that a violation of a statute constitutes negligence per se. The effect of this instruction is that the plaintiff may not have to prove that the contractor was negligent in grading and compacting, but may need to prove only that no licensed grading contractor was on the job. A good plaintiff's lawyer will then argue to the jury that if the contractor cannot be trusted to use a licensed grading contractor for this important part of the work, many other things must be wrong with the project. From that point on, the contractor is tainted in the eyes of the jury. This is a good example of the way in which the temptation to use a shortcut can have disastrous results in a lawsuit.

Poorly compacted soils can cause dramatic damage to a structure. Figure 5-1 depicts the interior of a single family residence that was constructed on approximately 65 feet of poorly compacted fill. This residence had a post-tensioned slab-on-grade foundation that was found to have cracked directly across the entire house. In short, the house had broken in two. The photographs show extensive cracks in the interior drywall, some of which are being monitored by "crack gauges," devices that are used to show movement over time.

The same poorly compacted soil at this residence caused the concrete flatwork to crack, separate, and fail, as shown in Figure 5-2.

Poorly compacted soil will also manifest itself in the settlement of the structure, as shown in cracks to the exterior plaster system. Figure 5-3 shows extensive cracks in a concrete wall due to settlement from poor compaction.

How to Avoid these Defects

- Confirm that the soils report has verified the height of the water table and that the soil is not saturated. In a large area of land, verify that tests have been made in all areas of the project to determine the water table and the moisture content of the soil.

- Verify the existence of expansive soils and obtain specific recommendations from the soils engineer—consider removing and replacing the expansive soil. Scrupulously follow the soils engineer's recommendations for foundation systems.

Figure 5-1

Crack gauges measure the extensive cracks in the interior of this house

Figure 5-1
(*Continued*)

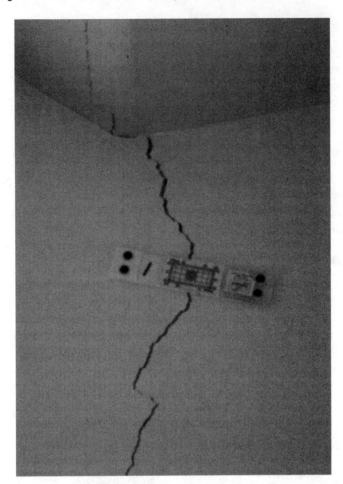

- With regard to compaction, use only licensed grading contractors, verify that the soil is free from organic material, verify that the soil is free from rocks and other foreign materials, verify that the grading contractor has followed the recommendations of the soils engineer with respect to moisture content and the thickness of lifts, and verify that the soils engineer has taken adequate soils samples to confirm compaction.
- Obtain specific recommendations from the soils engineer with regard to installation of the foundations and isolation of the foundation system from the saturated soils.
- Advise your client about the effects of expansive soils.
- Use only licensed grading subcontractors or have your own grading license.
- Strictly comply with the requirements of the soils engineer.
- Obtain and keep a copy of the soils engineer's inspection reports showing the compaction of the soil.
- Photograph the condition of the soil immediately before the placement of rebar or the pouring of concrete.

Figure 5-2
Poorly compacted soil caused cracks in concrete and in the mortar joints of a concrete block wall

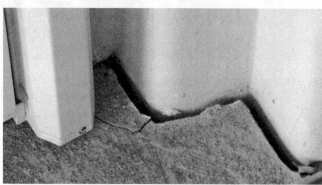

Figure 5-3

Settlement from poor compaction caused extensive damage to the interior of this home

Defects in the Installation of Foundation Systems

Recall that foundation systems are of two basic types: slab on grade and raised foundations (foundations supported by caissons and grade beams are discussed separately). The typical construction defects for each of these systems relate to reinforcing of the concrete, providing openings for utility connections, placing anchor bolts and hold downs, and protecting the foundations from water, including the ventilation and water-proofing of crawl spaces.

Improper Reinforcing of Concrete

In slab on grade foundations, the concrete must be reinforced so that it does not crack and so that it better resists movement of soil. Two kinds of reinforcing materials are used to reinforce the concrete in slabs: welded wire mesh (a grid-like product of thick wires welded together to form squares) and rebar. Either material can be used, but both must be placed in the middle of the concrete slab. To accomplish this, the reinforcing material must be held up off the ground, using small supports made of metal or plastic, called *dobies* or *chairs or small pieces of concrete.*

The typical defect occurs when the concrete subcontractor does not support the reinforcing material and the reinforcing material ends up at the bottom of the concrete slab, not in the middle. This usually happens because it is time-consuming to place the small supports under the reinforcing material and because it is difficult to walk around the area when the material is off the ground. When the reinforcement is placed in the bottom of the slab and not in the middle, this weakens the slab and makes it susceptible to cracking and failure. Reinforcement that is placed too close to the surface of concrete (less than 2 inches from the surface) is also more susceptible to rusting and premature deterioration.

Failure to Prepare for Utilities

When a slab on grade foundation is used, the utility connections—i.e., electrical lines, plumbing lines, sewer lines, and hvac ductwork—have to be placed in the ground under the slab and must penetrate the slab to enter the structure of the house. To accomplish this, openings are formed in the slab before it is poured. This is called *blocking, sleeving,* or *canning.* The locations and sizes of each of these openings must be carefully determined so that when the utility lines are brought into the house, they arrive in the right place and there is sufficient room for the various lines and ducts.

Failure to Install Foundation Bolts Properly

In raised foundations (as in slab on grade foundations), a series of anchor bolts and bolts for hold-downs must be installed when the concrete is still wet. Typical construction defects in the installation of these devices follow:

- Failure to install anchor bolts using the spacing required by the structural engineer—i.e., anchor bolts that are too far apart.
- Failure to install anchor bolts at the correct height, so that a washer and nut assembly can be placed on top of the sill plate and adequately secured.
- Failure to place the anchor bolts so that they are plumb (making it difficult to secure the washer and nut assembly properly on top of the sill plate).
- Failure to place anchor bolts a sufficient distance from the end of a sill plate or corner of the structure. (anchor bolts that are placed too close to the ends of sill plates or corners of the structure are not approved by the structural engineer because there is not a sufficient amount of wood to be restrained by the anchor bolt assembly.)

Crawl Spaces

The crawl space lies between the top of grade and the bottom of the floor joists in a raised foundation system—it is the space under the house where a worker can crawl around to install or service various assemblies.

Much has been written about crawl spaces and crawl space ventilation in recent years, as concerns about mold have moved into the public consciousness, and much attention has been given to the subject by contractors, architects, and waterproofing consultants. It is also true that no consensus exists on how best to treat crawl spaces to avoid excessive moisture that may lead to fungal contamination. That said, the evolving knowledge on the subject suggests that certain construction methods be implemented, and some construction methods assuredly fall into the "defect" category.

A starting point is the uniform building code (UBC), which requires that the distance between the top of grade and the bottom of the floor joist be not less than 18 inches. The UBC also requires that crawl spaces have ventilation openings in the walls of the crawl space of not less than 1 square foot for every 100 square feet of floor area of the crawl space.

Like all provisions of the UBC, the provisions calling for 18 inch crawl spaces and crawl space ventilation are minimum requirements, and they can be greatly improved by better construction methods. Because of the enormous concern about mold, and because of the increasing complexity of homes and the attendant need for more systems (and more wires, ducts, and assemblies), all of which seem to require ever-increasing servicing, I recommend crawl spaces of at least 36 inches, and if possible, 48 inches (4 feet). A tall crawl space will not only allow construction to proceed more quickly, it will result in a better and cleaner installation of systems and greater ease in servicing them.

With regard to ventilation of crawl spaces, this is a more difficult area. One school of thought proposes that crawl spaces be made part of the "conditioned" air space of the building. In this kind of installation, the crawl space is carefully sealed, meaning that the ground is sealed off from the structure, the walls of the crawl space are insulated and heating and air conditioning ducts are run into the crawl space so that it is heated in the winter and cooled in the summer. Another approach uses the idea of sealing the crawl space but not including it in the conditioned air space, leaving the space without ventilation. A third approach uses the idea of sealing off the ground but ventilating the space.

I would make the following observations about these different approaches. If a crawl space is to be sealed, whether or not it is ventilated, the work must be done with extreme care, because any gap in the assembly that allows moisture into the crawl space without a way for the moisture to escape could prove problematic and difficult. At the same time, if a crawl space is to be ventilated in a more traditional way, it must be well ventilated to avoid the accumulation of moisture and resultant possibility of mold growth. In either instance, preventing water from reaching the crawl space is an essential element in a successful installation.

It is in this latter regard that most construction defects arise—i.e., the failure to ventilate crawl spaces adequately. In many instances, the code minimum requirement for ventilation is not met. In some other instances, the code minimum is met, but because it is simply a minimum, it does not perform adequately.

In many newer homes, designers wish to achieve the look that the home is level with grade. To ventilate crawl spaces in this kind of configuration, a well must be built adjacent to the outside wall of the crawl space and a vent installed at the bottom of the wall inside the well. This technically satisfies the building code requirement, but in practicality it performs rather poorly, because there is no direct path for cross-ventilation of the crawl space. In addition, leaves and other debris tend to accumulate at the bottom of these wells, covering the already limited ventilation space.

Related is the issue of drainage from the site around the foundation. Even a properly ventilated subfloor area may accumulate too much moisture (leading to the growth of mold) if the drainage around the foundation is not adequate. It was thought for many years that raised foundation walls did not require waterproofing below grade. More recent construction techniques include the complete waterproofing (not merely damp-proofing) of foundation walls and the construction of perforated drain systems around foundation walls to prevent the intrusion of water into subfloor spaces. Waterproofing can be accomplished below grade using any of the standard methods—i.e, sheet applied systems, liquid applied systems, or bentonite clay panels, combined with a perforated drain system to carry water away from the foundations.

Careful attention to this issue at the planning stage can avoid the defect claim after the job is finished. In addition, sloping of the soils away from the foundation walls and placing a layer of plastic under the surface of the ground to act as a drainage plane (if planting is to be installed around the foundations, holes will have to be cut into the plastic sheeting for root growth), will reduce the amount of water reaching the foundation walls.

Area drains around the structure to reduce the amount of standing water are also important in the overall performance of the structure over time. Careful attention to area drainage during the early stages of the project is essential, because area drains need to be connected to drain lines that are buried underground, and the sloping of these drain lines can be affected by underground utility lines, trees, walls, and other objects.

Foundation Waterproofing

While considerable debate continues in the construction community about the need to waterproof raised foundation walls below grade, there is no debate on the subject of the need to protect slab on grade foundations from water and water vapor.

Recall that all concrete is porous. (Some concrete mixes resist water much better than others, and concrete can be sealed and otherwise treated to resist water, but a conventional concrete slab is permeable to water and water vapor and must be protected against water intrusion.) When a concrete slab is placed on grade, unless a water or vapor barrier is installed, water or water vapor can permeate the concrete, leading to moisture inside the structure of the house.

It is critical at the outset to determine whether a water barrier or a vapor barrier will be required. In nearly all conventional situations, a vapor barrier is all that is necessary. In areas where the water table is high, or where a great deal of rain falls or other surface groundwater is present, a water barrier may be required. The materials and techniques of installation differ substantially for water barriers and vapor barriers.

Most construction defects arise from the installation of vapor barriers, because water barriers are more rarely installed. A vapor barrier is customarily a layer of thin plastic, usually called Visqueen (which is actually a trademarked name), sandwiched between two layers of sand, each of which is 2 inches thick. The typical defects are failure to install any sort of vapor barrier; installing a vapor barrier incorrectly, resulting in gaps that provide a path for the migration of water vapor; and installing a vapor barrier correctly but failing to protect it adequately after installation, leading to rips and tears and a path for water vapor transmission.

To avoid the defect, do the following:

- Carefully review the plans to determine the type of reinforcing specified for the slab on grade foundation.
- Install chairs or dobies to insure that the reinforcing wire or rebar is placed in the middle of the slab.
- Carefully lay out all utility locations before pouring the slab and verify their location and size.
- Carefully review the plans to determine the spacing and location of anchor bolts and bolts for hold-downs, install all bolts at the correct height and in plumb.
- Verify that code minimums have been achieved for crawl spaces and recommend that a 36 to 48 inch crawl space be constructed.
- Provide adequate crawl space ventilation and recommend the use of a powered ventilation system.
- Recommend the installation of below grade waterproofing on raised foundation stem walls and the installation of perforated drains around stem walls.
- Verify the amount of groundwater (review the soils and geological reports carefully) and carefully install vapor or water barriers so that they are monolithic and well protected from damage until the installation of the foundation slab, including the installation of two layers of 2 inch thick sand.
- Advise the client in writing against design concepts that will prevent good ventilation.
- Provide a maintenance schedule to keep vents clear.

DEFECTS IN CIVIL ENGINEERING AND SITE DRAINAGE

In case it hasn't become clear so far, water intrusion into homes is a major, if not the number 1, cause of construction defect claims and is to be avoided at all costs. Every expert will tell you that drainage of surface water away from homes is a critical element in resisting water intrusion and maintaining a home in good condition.

A substantial amount of water can be carried away from a home through surface, or sheet, drainage if the area around the home is graded so that the land slopes away from the structure. In areas where it is not possible to grade away from the home, an alternative drainage system must be provided—either surface drains (area drains) or underground drainage systems such as perforated drains.

Concrete flatwork is often constructed around homes, in the form of patios and landings. These installations are usually immediately adjacent to doors, either conventional swinging doors or sliding patio type doors. When patios, decks, and landings are not sloped to drain away from the home, and they're not sloped away from the doors leading into the home, the builder has created a serious potential for water intrusion.

Thus, defects in this area tend to fall into three major categories: the land around the home is not graded to slope to drain away from the home, no drainage system is installed, and the hardscape, such as patios and landings, are installed with reverse slope—i.e., they slope toward the house and not away from it.

With regard to perforated drainage systems, even when these systems are installed, it is quite common to see them fail because of two factors: an insufficient amount of crushed rock below, around, and on top of the perforated pipe; and failure to install filter fabric around the crushed rock and perforated drain. The crushed rock around a perforated drain functions to bring water from the surrounding area rapidly. The faster the water can drain from the surrounding area into the drain, the less hydrostatic pressure will be present against the structure of the building (or retaining wall). The less hydrostatic pressure, the less chance for a penetration of the waterproofing system and water intrusion. Many contractors install only the minimum amount of crushed rock (a few inches below the perforated pipe and perhaps 12 to 18 inches above). A better installation uses as much crushed rock above the pipe as possible—such as crushed rock to within a foot or so of grade.

Similarly, all the crushed rock in the world won't be effective if it becomes clogged with mud and silt. To avoid this, the filter fabric must be placed so that mud and silt cannot intrude into the crushed rock installation.

Here's how you can avoid the defect:

- Consult with the civil engineer, architect, or other design professional to determine appropriate drainage from the site.
- Counsel the homeowner to install appropriate site drainage in the form of area drains or underground drainage systems.

- Verify that all hardscape drains away from the building structure.
- Excavate to the bottom of the footing to install perforated pipe. Install the perforated pipe with the drainage holes on the bottom, not the top. Verify that the pipe is sloped to drain. Prior to installing the perforated pipe, install an adequate layer (at least 6") of 3/4" crushed rock. Install at least two feet of crushed rock over and around the perforated pipe. Install filter fabric to prevent the intrusion of mud and silt.

DEFECTS IN FRAMING

Construction defects in framing generally are limited to a small number of items, and most arise from the defective installation of shear panels. Other framing defects commonly seen are lack of horizontal fire stopping, walls that are out of plumb or out of plane (i.e., not straight along the length of the wall or bowed out or in vertically from floor to ceiling), lack of full height blocking (explained later), and reverse or otherwise improper sloping of balcony deck.

Defects in Installation of Shear Panels

Shear panels deserve special attention because they are an essential aspect of the overall ability of a structure to resist lateral movement and to resist wind loads. As a result, defectively installed shear panels are "life safety" issues that can result in the collapse of a structure and serious injury or death to the occupants. In addition, it is expensive to determine whether a shear panel was correctly installed after a house is built, because it requires the removal of finished surfaces such as stucco, and the cost of repairing defective shear panels is also substantial for the same reason. Therefore, it is critically important to make sure that the installation of shear panels is correctly done in the first instance.

Claims of defects for shear walls are as follows:

- Incorrect thickness of plywood or oriented strand board material
- Shear walls installed using the wrong type of nails
- Nailing pattern not followed
- Nails driven too close to edge of plywood
- Nails overdriven, puncturing the skin of the plywood

Most commonly, shear wall nailing problems evidence themselves through overdriven nails. Generally speaking, nails should be three times longer than the thickness of the wood that the nail is intended to hold, because two thirds of the nail should be anchored in the supporting piece of wood. Thus, for a shear panel that is 1/2 inch thick, a nail of at least $1\frac{1}{2}$ inches should be used. One-half inch of the nail will be driven through the shear panel and 1 inch will be embedded into the framing behind the shear panel. Virtually all shear walls are installed using pneumatic nail guns, each of which can be set to drive the nail to a certain depth in the material. The problem is that the installer may not adjust the gun, the result being that the nail head is driven through the top surface of the plywood shear panel. From the structural engineering perspective, any nail driven through the skin (the top surface of the plywood) is a nail that has less structural value than a properly driven nail, for two reasons: driving the nail through the skin of the shear panel reduces the ability of the shear panel to resist tearing forces in a seismic event; and driving the nail through the skin reduces the amount of friction on the nail from the plywood panel—i.e., the nail is surrounded by less than 1/2 inch of plywood.

Similarly, nailing patterns are of high concern to engineers. Basically speaking, a shear panel is a device that functions to resist lateral forces. In essence, it makes a wall stiffer. To determine how strong the shear panel needs to be, the structural engineer determines the mass of the building and the theoretical lateral force that will be applied to that mass. He or she then determines how much resistance the shear panel must provide. The resistance is derived from the strength of the wood panel and the manner in which it is attached to the wood framing. The level of resistance is thus a function of the thickness and type of wood used for the panel, and the type and spacing of the fasteners.

If the structural engineer has specified that the panel is to be nailed every 3 inches on the perimeter, and the installer has nailed the panel every 6 inches, the structural engineer will have a legitimate complaint that the ability of the shear panel to resist against lateral forces has been compromised.

Similarly, if nails are overdriven or driven too close to the edge of the shear panel, the structural engineer will take the position that these nails do not serve their intended function to attach the panel adequately to the framing of the home. Figure 5-4 shows typical overdriven nails.

Figure 5-5 shows properly driven nails in a plywood shear panel.

Lastly, if the wrong size nail is used, the structural engineer will take the position that the holding power of the undersized nail does not comply with the structural calculations and will require that the shear panel be removed and replaced with properly sized nails. Sometimes the correct length of nail is used, but the wrong kind of nail is used. Framing of shear panels requires the use of "common" nails. Many times, "box" nails are substituted for common nails. The size of a nail is described using the word *penny*, which is abbreviated as *d*. For example, an 8d nail (8 penny) is $2\frac{1}{2}$" long. However, although an 8d common nail and an 8d box nail are both the same length, the box nail is thinner than the common nail. Because it is thinner, it does not grip the wood around it with the same force and it is less resistant to bending and tearing forces. Thus, an 8d common nail would be appropriate to install a shear panel, but the use of an 8d box nail would be a construction defect. It should be easy to see from this explanation that it is simple to avoid problems with shear panel installation, and the correct installation should add literally nothing to the cost of the project.

Here's how to avoid the defect:

- Confirm that the plans specify the thickness and type of plywood or oriented strand board (OSB) to be used for shear walls.

- Strictly comply with the plans.
- Do not substitute materials without written consent of the structural engineer.
- Instruct the framing subcontractor that all nailing patterns, types of nails, and installation of nails must strictly comply with the plans and specifications and obtain and maintain in your files a document signed by the framing subcontractor acknowledging that he or she has been so advised.
- Prior to the installation of the first shear panel, verify that the nailing guns have been properly set so that nails are not overdriven.

Figure 5-4

Improper shear wall nailing

Figure 5-4
(*Continued*)

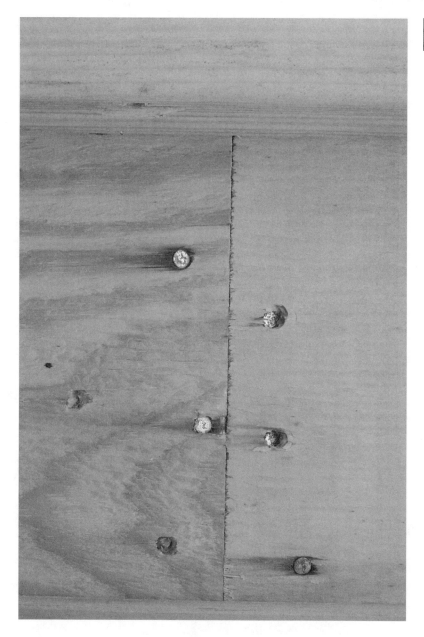

- Verify that the correct type and size of nail is being used.
- Inspect daily for compliance with nailing patterns and driving of nails.
- Replace any shear panel that does not strictly comply.
- Place numbers on the plans to correlate to the shear panels. Number the shear panels with spray paint or marker. Photograph each shear panel so that all nails are visible. Maintain the numbered plans and photographs in your job file. Have the structural engineer sign off after his or her inspection.

Figure 5-5

Proper shear wall nailing

Defects from Omission of Fire Stopping

Fire blocking in framing is similar in many ways to shear walls because it is also a life-safety issue, it requires a great deal of money to investigate and repair, and it is quite simple to avoid in the construction process.

The UBC requires the installation of horizontal fire blocking in stud cavities to provide a one-hour fire rating. This construction defect has only one cause: the failure to install the blocking. This one is simple.

Here's how to avoid the defect:

- Install the fire blocking.
- Photograph the walls before they are closed up.

Defects from Omission of Full-Height Blocking

The intersection of the roof rafters and top plate typically requires the installation of a block between the rafters as a brace to keep the rafters from moving from side to side and to provide a load path from the roof sheathing downwards into the walls. In the typical installation, to create this block, the framer must rip and angle a piece of 2 × 10 or 2 × 12 to fit snugly between the top plate and the bottom of the roof sheathing. To save time, the framer may use a piece of dimensional lumber without ripping it down to size or providing the angle that reflects the slope of the rafter. The result is a gap between the top of the block and roof sheathing. As a result, there is no direct connection between the sheathing and the block (because there is a physical space between the bottom of the roof sheathing and the top of the blocking), creating a discontinuity in the load path from the roof diaphragm to the walls of the structure. A discontinuity in the load path can cause the roof diaphragm to shear away from the top plates and seriously damage the structure.

The photograph shown in Figure 5-6, taken from inside an attic space looking out toward the outside wall of a single family residence, shows the correct installation of blocking. Note that the blocks are flush with the bottom of the roof sheathing material.

Figure 5-6
Correct installation of blocking

Although this is a simple defect to avoid, it occurs in a surprising number of structures. Additionally, unless access to the area is available to make repairs, the installation of full height blocking after the roof is installed can make the repair of this defect extremely expensive. Since a discontinuity in the load path creates a serious structural defect in the building, it can be argued (and usually is argued successfully) that this is a life-safety issue, for which a repair is absolutely necessary.

Avoiding the defect is simple (perhaps the simplest of all of the construction defects discussed in this book):

- Measure the distance between the top of the plate and the top of the rafter.
- Calculate the angle created by the pitch/slope of the rafter.
- Transfer the angle to the dimensional lumber to be used to construct the block.
- Verify that the height of the block is such that the block will end up at the same height as the top of the rafter.
- Carefully cut the angle to match the pitch of the rafter.
- Install the full height and angled blocking.

Defects from Out of Plumb and Out of Plane Walls

Framed walls are, in a perfect world, plumb and flat. In other words, the walls are perfectly straight up and down and flat along the wall surface. In the real world, walls are seldom completely flat, although they should be plumb. Factors that affect the flatness of a wall include the height of the wall, the quality of the lumber, and the type of lumber.

These types of claims usually arise after the drywall has been installed and items such as mirrors are being installed. The glass subcontractor arrives to install a large mirror and discovers that the wall is out of plane to the extent that the mirror cannot be attached to the wall. This is an expensive repair for the contractor.

Some degree of variance in plumb and plane is to be expected in light wood frame construction, and some of it results from the quality of the framing material. In this regard, it helps to know what the standards are.

Although there is no uniform or absolute standard for tolerances in rough framing, generally, walls should be plumb within 1/4 inch for any 32 inches in vertical measurement. For walls that will be clad with gypsum wallboard, a tolerance of 1/4 inch in 10 vertical feet is recommended.

Many problems in this area can be avoided by informing clients that because the lumber available today is of poorer quality than years past, some warping is to be expected. Good client communication would also require explaining to clients that for additional cost, premium lumber (lumber that has been kiln dried beyond normal standards and that has fewer knots, wanes, and warps) could be used to minimize these problems, or that some consideration could be given to the use of steel studs. Steel studs will eliminate out-of-plane walls if correctly installed, but most framing crews will not install a combination of conventional wood framing and steel in-fill stud walls (the additional structural concerns with steel in-fill walls require special structural engineering).

One way to minimize or eliminate claims for out-of-plumb and out-of-plane walls is the use of engineered lumber. The framing materials generally available today are of poor quality, leading to considerable twisting and bowing in otherwise well-built walls. Engineered lumber is made of small pieces of wood that are glued together under heat and pressure. The advantage of using engineered lumber is that it is straight and true, and it tends to stay that way to a much higher degree than conventional framing lumber. Engineered lumber is sold under various trade names, each with its own distinctive characteristics. One disadvantage to using engineered lumber is price. It is significantly more expensive than conventional lumber. Another issue is that, generally speaking, engineered lumber must be connected only to other engineered lumber. For example, if engineered floor joists are used, the wood that is used to block them at the ends (the rim joist) should also be engineered lumber. This can add to the expense of the project. Finally, it could be argued that the use of engineered lumber has some environmental benefits, because the engineered lumber is made from what is basically scrap. This kind of lumber can also be manufactured from farmed trees and not old-growth timber.

After the framing is installed, it can be checked using a long straightedge or laser level. The framing contractor can correct bowing or other out-of-plane problems by planing the surface of the studs, a process called *staightedging*.

As an alternative as noted above, consider the use of steel stud in-fill walls ("tin can" studs). Particularly in areas where other parts of the structure provide structural support, steel studs are an attractive alternative and if installed correctly provide walls of near perfect flatness. It is possible to construct complex shapes from steel studs as well. Figure 5-7 shows examples of steel stud construction in the basement and garage of a large custom home.

Here's how to avoid the defect:

- Closely inspect the work of the framing contractor with a laser device or straightedge to ensure that the walls are within acceptable tolerances for plane and plumb.
- Require the framing contractor to correct deficiencies in the framing by planing the framing before the installation of drywall.

Defects Arising from Balcony Deck Framing

Balcony decks are designed to drain in two ways: by sheet flow toward a scupper, or by drainage to an area or deck drain, with a secondary drainage device such as scupper provided in the event that the primary drain is blocked. For either system to perform properly, the framing of the balcony deck must be sloped toward the drainage device. In a design using scuppers as the primary drainage device, this means that the balcony deck framing must slope away from the building facade and toward the scupper. When deck drains are used, the framing must slope toward the deck drains.

Figure 5-7

Steel stud
construction in a
basement and
garage

Figure 5-7
(*Continued*)

In far too many cases, balcony deck framing is constructed flat or with a slope toward the building structure (reverse sloping). This will result in drainage toward the building, ultimately leading to water intrusion, usually through or under the door thresholds or door sill pans.

Avoiding the defect is simple: verify that the framing has been sloped correctly toward the primary drainage device. A more complete discussion of balcony related defects is set forth later in the chapter.

A related and also common defect is an insufficient slope. In this circumstance, the framing is sloped, but the amount of slope is not sufficient. As a result, water will tend to pool, or pond, on the surface of the balcony deck and on the waterproofing material below the deck surface. Ponding water will gradually cause the waterproofing system to fail and it is a primary source of water intrusion into the building structure.

Here's how to avoid the defect:

- Lay out the direction of slope.
- Verify that sufficient freeboard exists between the door threshold and the outer edge of the balcony to provide the code required degree of slope.
- Verify by use of line or level that the code-required amount of slope, usually 1/4 inch per foot, has been provided.

DEFECTS IN PLUMBING

The usual defects in plumbing arise from the following areas: failure to ream the cut ends of copper pipe; failure to isolate pipe from the framing of the building, leading to noise; failure to install dielectric fittings where copper pipe is connected to galvanized pipe; and failure to install temperature pressure relief valves or install them in a proper manner. Of all these issues, by far the most important is the reaming of cut ends of copper pipe.

Failure to Ream Cut Ends of Copper Pipe

Although the uniform plumbing code (UPC) makes it clear that the cut ends of copper pipe are to be reamed, there seems to be a general lack of understanding in the community of plumbing installers as to the reason for this requirement and the consequences of not following this important code specification.

Copper pipe is designed so that water will flow in a *laminar* fashion. This means that the water flows through the pipe in an uninterrupted, smooth fashion, with no roiling in the stream of water. As the water flows through the pipe, the water deposits a thin film on the inside of the copper. This thin film prevents the copper from eroding over time.

When the end of a copper pipe is cut, the cutting tool does two things: it crimps the end slightly, narrowing the inside diameter of the pipe; and it creates a burr on the inside of the pipe at the end where the cut occurred. The UPC requires the contractor to take a small deburring tool, or a knife, and de-burr, or ream, the inside of the cut end of the pipe. This takes only a moment, and it does two things: it opens the pipe to the original inside diameter, and it removes the burr.

If the burr is left on the inside of the pipe, and the pipe is then connected to a coupling, an elbow, or other fitting, the burr interrupts the laminar flow of the water. When the laminar flow is interrupted, instead of the water flowing in one direction through the pipe, the burr creates a sideways current about 6 inches downstream from the burr. Water flows through copper supply lines at about 6 feet per second (fps). When the burr creates a current, the water flows directly against the walls of the pipe—that is, sideways and perpendicular to the overall direction of the water flow. This perpendicular or sideways current causes the water to strike the inside of the walls of the pipe at nearly 100 fps.

The force of these sideways currents wears off the thin deposit on the inside walls of the pipe. When the protective coating is worn away, the raw copper is exposed to the strong current. Eventually, the current will cause the copper to erode, and a pinhole leak will result. This phenomenon, well-recognized in copper piping systems for more than 70 years, is called *erosion corrosion* or *cavitation erosion*, and the result is pinhole leaking throughout the system.

When this occurs, the only fix for the problem is the total removal of the system and total replacement. Depending on water conditions, the type of copper pipe used, the heat of the water, whether a circulating pump is used, and the extent of the burrs, it may take as little as one year or as long as 20 years for pinhole leaks to develop. Eventually, however, pinhole leaks will develop in copper systems where the pipes have not been reamed.

This is perhaps the simplest potential construction defect to avoid. All plumbers working on the job should be instructed to ream the cut ends of copper pipe. This simple and quick act will avoid the defect in its entirety. A good "spot check" can be made by looking for copper shavings in the area that the plumber has just completed.

Here's how to avoid the defect:

- Ream the cut ends of all copper pipe.
- Obtain and maintain in your file a document from the plumbing subcontractor in which the plumbing subcontractor acknowledges having been instructed to ream the cut ends of all copper pipe.
- Consider including in your contract with the plumbing subcontractor a provision that entitles you to make a random inspection of the copper pipe to determine whether the cut ends have been reamed, and if you find a non-reamed pipe, to require the removal and re-installation of the copper pipe system at the subcontractor's expense.

Failure to Isolate Pipe from Framing

Plumbing supply lines and drain, waste, and vent lines all have the potential to create noise in the building structure. Drain lines in particular, and especially those that are constructed from plastic ABS pipe (as opposed to cast iron pipe) have the potential to create noise. While noise is a problem in any dwelling, it is a particular source of construction defect litigation in multi-family housing, such as condominiums. Many condominium construction defect lawsuits start with complaints about noise.

Noise from plumbing lines can be reduced, or in many cases eliminated, by proper isolation of the pipe from framing members. When a pipe is directly attached to framing, without any isolation device, the vibration from the pipe is transmitted to the framing member. The vibration then travels from the framing member to the drywall, which in many instances acts in much the same way as a speaker cone, directing the sound into the room.

Pipes can be isolated from framing in several ways, among them by injecting expandable foam into the small space between the pipe and the framing, or using plastic attachment devices that isolate the pipe from the adjacent framing member. I advise that you use the services of an acoustical engineer to obtain specific protocols for the isolation of plumbing lines and the reduction of noise.

Figure 5-8 shows a careful installation of a large cast iron waste line and copper supply lines that have been isolated from the framing using felt wraps.

Here's how to avoid the defect:

- Verify that all supply, drain, waste, and vent lines have been properly isolated from the building structure.
- Advise customers in writing of the sound-reducing qualities of cast iron pipe and recommend that an acoustical engineer provide a protocol for the installation of the plumbing lines.

Figure 5-8
Proper methods of isolating pipes from the framing

Figure 5-8
(*Continued*)

Failure to Install Dielectric Fittings

When copper and galvanized pipe are in direct contact with one another, galvanic corrosion will occur, resulting in the deterioration of the connection between the dissimilar types of pipe material. Eventually, the connection will fail and a leak will result.

This is a simple problem to avoid. In instances where copper pipe must be connected to galvanized pipe, a dielectric union or fitting must be used. This device separates the copper from the galvanized pipe and prevents galvanic corrosion.

Failure to Install Temperature Pressure Relief Valves

Hot water heaters require the installation of temperature pressure relief (TPR) valves. A TPR valve is designed to reduce excessive pressure in a hot water heater in the event that the heater malfunctions and builds up too much pressure.

Omission of the TPR valve happens surprisingly often, even though the TPR valve is required by code and is highly visible—in many instances it simply gets missed by building inspectors. Incorrect placement of the TPR valve is also common. A TPR valve should be placed less than 6 inches above the finished floor surface so that it does not injure a person standing in the area in the event that hot water or steam is expelled from the valve without warning. In many cases, the valve is installed on the hot water heating appliance at eye level.

To avoid the defect, verify that the TPR valve is installed and that the installation is at the correct height.

Defects in Electrical Systems

Relatively speaking, few construction defect cases start with problems in electrical systems. Various theories have been expressed as to why so many more defects occur in other building components and installations, among them that electricians are better trained than many other trades. Regardless of the cause, the typical defects are easy to avoid and are relatively simple in nature:

- Failure to identify and label circuit breaker boxes adequately to inform the owner of what circuit breaker regulates what circuit.
- Failure to install receptacles flush with finished wall surfaces and to secure them to the J box in which they are installed.
- Failure to install receptacles at a uniform height.
- Defects in the penetration of structural framing members.
- Defects in the penetration of the building envelope.

With regard to labeling, the National Electrical Code requires that circuit breakers be clearly labeled so that the building owner can easily identify a particular circuit breaker in the event that service is required for a receptacle or switch.

With regard to receptacles, switches and receptacles are attached to a metal or plastic box, which, in turn, is nailed or screwed to the framing of the building. Receptacles and switches need to be placed so that they are at the proper distance from the finished wall surface. In many homes, particularly higher-end construction, the placement of receptacles is a significant part of the overall design. In those circumstances, uniform placement of receptacles is important.

Defects in the penetration of structural members arise from the failure to observe code requirements that are designed to insure that structural framing members are not adversely affected by penetrations for electrical conduit. The code restricts the size and placement of holes in joists and plates. Holes that are too large or placed too close to the edge of joists and plates impact upon the structural integrity of those members.

Electrical contractors often are required to make openings in the outside walls of the building. Waterproofing these openings can be difficult if the openings are cut after the application of the exterior finish, such as stucco, and penetrations of the building envelope by electrical contractors are a prime source of water intrusion–related construction defects.

Here's how to avoid the defect:

- Clearly label all circuit breakers.
- Verify that all switches and receptacles are set properly.
- Verify the uniform installation of receptacles during installation.
- Verify that joists and plates have not been penetrated in violation of applicable code requirements.
- Verify that penetrations of the building envelope have been appropriately waterproofed.

DEFECTS IN WATERPROOFING THE BUILDING ENVELOPE

Construction lawyers all agree on one thing: virtually every construction defect case starts with water intrusion. Water intrusion not only is the trigger for almost every case, but it causes extensive and expensive damage. Now that mold has become a part of the public consciousness, water intrusion is even more important.

If you take away only one recommendation from this book, that recommendation is that you take extreme care to waterproof the building envelope. This alone will help substantially in avoiding lawsuits.

As a general proposition, because this area is so vitally important, I strongly recommend that all contractors require the owner to hire a waterproofing consultant to review the plans and specifications and to make specific recommendations for below and above grade waterproofing. The simple fact is that architects generally lack the necessary practical experience and expertise to devise adequate waterproofing details. Waterproofing experts can and will provide plan details, specifications, and recommendations for installation of waterproofing systems and materials. When you obtain specific recommendations from waterproofing consultants, and you follow those recommendations to the letter, you immediately reduce not only the chance of a problem, but also your potential liability.

Keep in mind at all times that *stucco is not waterproof.* Any penetration of the building envelope that is not properly waterproofed will leak, even if covered with stucco. Remembering this simple fact will eliminate many of the most common construction defects leading to water intrusion.

Below Grade Water Issues

Below grade water issues manifest themselves differently in single family homes than in multi-family structures. Single family home below grade water issues typically involve the failure to install a vapor barrier or waterproof membrane between grade and the concrete foundation slab in the case of slab on grade construction—as noted earlier in the chapter, failure to ventilate subfloor spaces adequately and failure to waterproof areas adequately where grade is above the level of the sill plate. In some instances, problems also occur where retaining walls have not been waterproofed or the civil engineering has improperly allowed drainage to flow toward the structure.

In multi-family structures, which often involve the use of subterranean or semi-subterranean parking structures, water intrusion from improperly waterproofed masonry walls or poured concrete walls is common. Because the inside of these walls is visible from the parking structure, and efflorescence almost always results from the water seepage, owners become immediately aware of the problem. Any leak into the interior of a subterranean garage will have the same effect, so special attention must also be paid to waterproofing around windows in basement/garage areas. Figure 5-9 depicts water intrusion in a subterranean garage in through a poorly waterproofed concrete masonry unit (CMU) wall and through a poorly waterproofed window.

Subterranean garage walls are typically not waterproofed because the project may be an in-fill type project that has building on both sides. As a result, there isn't sufficient room to excavate behind the newly constructed masonry walls to install waterproofing behind the walls (called a *positive side* waterproofing system). Either no waterproofing is

Figure 5-9

Water intrusion through a poorly waterproofed CMU wall and window

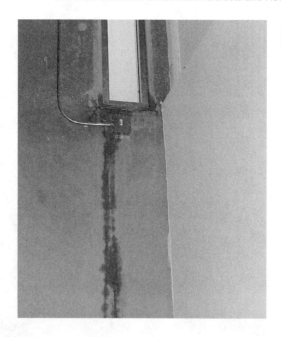

Figure 5-9
(*Continued*)

installed, or a waterproofing system is installed on the inside face of the walls (a *negative side* waterproofing system). Negative side waterproofing systems are problematic and generally perform poorly over time.

I recommend that contractors refuse to install any below grade structure of any significance—i.e., a subterranean garage space or a basement—for which no positive-side waterproofing has been specified. It is my experience with these structures that they simply cannot be waterproofed without a carefully designed and installed positive-side waterproofing system.

Even when positive-side waterproofing systems are installed, construction defects are not uncommon. They usually include the following:

- Failure to install a perforated drain system
- Installation of a perforated drain system, but at the incorrect elevation
- Improper installation of a perforated drain system
- Failure to install a monolithic positive-side waterproofing system
- Failure to protect the waterproofing system after installation, resulting in rips or tears in the system, allowing water penetration
- Failure to backfill properly against the waterproofed walls
- Failure to provide positive drainage away from the building structure

Figure 5-10 shows a retaining wall that was not waterproofed on the positive-side and another example showing the efflorescence that developed as a result of water infiltration through CMU walls. Because efflorescence is so visible, it serves as a red flag to

Figure 5-10

Damage incurred with poor positive-side waterproofing: a red flag to building owners

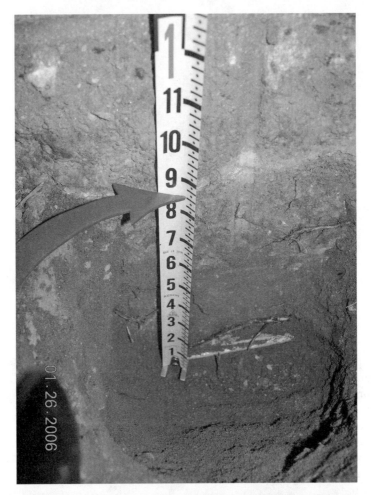

building owners and almost invariably results in investigation of construction defects and lawsuits.

Perforated drain systems are basic in operation, but like any installation, they require attention to detail. The most common defects (other than the complete omission of a perforated drain system) are perforated pipe placed upside down, perforated pipe placed at or above the level of the slab, an insufficient amount of crushed rock around and over the perforated pipe, and omission of filter fabric.

Perforated drain pipe should be placed with the holes down, always below the level of the foundation slab, with at least 6 to 12 inches of crushed rock, and filter fabric to prevent the intrusion of *fines* (fine grains of soil that can clog the crushed rock and block the flow of water) into the crushed rock. Where possible, it is better practice to backfill above the perforated drain with as much crushed rock as possible. In many instances, crushed rock can be used to backfill up to within a foot or so of grade, leaving only the amount of soil necessary for planting.

Below grade waterproofing systems are generally of three types: liquid applied, sheet membrane, and bentonite clay panels. Whatever system of waterproofing is used, the system must be installed and protected so that it presents a monolithic barrier to water. In the case of liquid applied membranes, the typical construction defects occur when the substrate (concrete block or poured in place concrete walls) is not clean and smooth prior to installation. Careful grinding of concrete walls is necessary to achieve a smooth surface prior to installation of any waterproofing system. In many instances, the liquid applied membrane is applied in a manner contrary to the manufacturer's recommendations and specifications, meaning that it is too thin or that some areas are not covered.

In instances where sheet applied membranes are used, the typical defects include the failure to prime the surface and failure to overlap the sheets adequately. Where bentonite clay panels are used, the system requires the installation of a termination bar at the top of the panels. Failure to install a termination bar will allow a gap between the top of the bentonite panel and the building surface, creating a path for water to intrude into the gap and run down the side of the building structure. The bentonite clay panels must adhere directly to the foundation with no foreign material between the panel and the foundation. In a recent case, bentonite clay panels were protected with Visqueen before installation. The installer failed to remove the Visqueen. The system failed because water intruded between the Visqueen and the foundation, rendering the entire waterproofing installation useless.

Failure to protect the system after installation is a common cause of problems. Newer systems use a tough plastic dimpled membrane on the outside of the waterproofing material both to protect the waterproofing membrane and direct water downward, thereby reducing hydrostatic pressure against the waterproofing membrane. Where no such device is specified, a protection board is almost always necessary to protect the membrane from abrasion and tearing from material in the backfill.

Figure 5-11

A well-installed waterproofing system

Figure 5-11 shows a well-installed sheet membrane waterproofing system, covered by a drain board and filter fabric.

Figure 5-12 shows another well-installed waterproofing system on the same residence. In this photograph, the earth has not yet been backfilled against the waterproofing system.

Backfill is an essential but often overlooked part of the underground waterproofing system. Backfill must be compacted to at least 90 percent relative compaction. Good compaction of backfill will significantly reduce the amount of water that will be introduced against the below grade structure.

Related to the issue of backfill is good positive drainage away from the building structure. The more water that can be diverted away from the area of the below-grade structure, the less likely it is that water will intrude through the below-grade waterproofing system.

Here's how to avoid the defect:

- Do not accept a job to install a below grade structure without a positive-side waterproofing requirement.
- Have the owner retain a waterproofing consultant.

Figure 5-12

Earth has not yet been backfilled against the waterproofing system

- Follow the expert's recommendations to the letter.
- Verify the proper installation of perforated drains. Proper installation includes adequate crushed rock, adequate slope to drain, installation of filter fabric, and backfill to 90 percent relative compaction.
- Photograph the progress of the work.
- Require the expert to inspect the work.
- Verify the proper installation of liquid or sheet applied membranes.
- Install a protection device for the waterproofing system.
- Properly compact backfill.
- Provide positive site drainage away from the below grade structure.

Moisture Proofing of Exterior Structures at Grade

Planters, support walls, garden walls, concrete flatwork, and other structures are often placed against the exteriors of wood frame structures with either no waterproofing or inadequate waterproofing. From the appearance of many of these installations, it would seem that many contractors believe that concrete is waterproof, because concrete seems to have been used as a waterproofing material. Suffice it to say that any structure that comes into contact with the building exterior above the plate line is a potential source for water intrusion into the wood structure. Therefore, particular care must be exercised to insure that waterproofing devices are installed. These can take the form of metal flashing or membranes of various kinds, usually liquid applied membranes or sheet applied membranes.

In areas of considerable rainfall, installation of waterproofing materials in the splash zone is recommended. The splash zone is the area in the first 18 inches above grade on exterior walls. The concept is that rainfall will hit the ground and splash upward against the building structure. Installation of extra waterproofing material in this zone can substantially add to the life of the building and protect it from serious water damage.

However, care must be exercised in the choice of waterproofing material. It is not uncommon to see the wrong kind of material used. For example, a leading supplier of waterproofing material and other building products makes two products—a rubberized asphalt and plastic peel-and-stick membrane waterproofing product. One of the products is intended for use in below grade applications and another is intended for above grade applications. If the below-grade product is used above grade, it will fail because it is not made to withstand the temperature extremes above grade, and it will delaminate from the building structure, allowing a path for water intrusion. A bit of care to insure that the proper material has been specified can save a great deal of time and aggravation down the road when the installation fails and water starts to intrude into the building.

Similarly, when below-grade materials are used under sheet metal above grade (such as coping caps), the extreme heat generated from sunlight on the metal surface will cause most of the materials to fail. Manufacturers make special materials for high-heat applications.

Recently, a number of products have appeared on the market that are used in the cold joint between the top of the concrete foundation slab and the bottom of the mud sill to seal the joint from air penetration and water intrusion. While I don't necessarily endorse any particular product, it would seem that particularly in areas of high moisture content or extreme weather conditions, thought should be given to the use of these products to provide a better structure.

In areas where planters, concrete terraces, support walls, and other structures that can transmit water to the building structure are placed directly against the building, careful waterproofing measures are essential. The installation of above-grade planters against buildings is a prime source of water intrusion and construction defect litigation. Particular care must be taken with planters to carefully and thoroughly waterproof.

Here's how to avoid the defect:

- Obtain specifications from the architect or a waterproofing consultants for the areas in question and keep the specifications in your job file.
- Strictly comply with the specifications.

- Strictly comply with the manufacturer's recommendations for preparation, priming, and installation.
- Do not substitute materials without written approval.
- Require the expert to inspect the work.
- Photograph the area before and after installation.
- Do not puncture, tear, or otherwise compromise the integrity of the materials.
- Protect the waterproofed area from damage from other construction activity.
- Flood or spray test the area before soil is backfilled or other structures built against the waterproofed building component.

Moisture Proofing of Retaining Walls

From the many claims in this area, it would appear that little thought is given to positive-side waterproofing of retaining walls or other masonry wall structures, because investigation of water seepage through those structures invariably reveals a complete lack of positive-side waterproofing material. In some rare instances, waterproofing material has been installed, but defectively.

The simple fact is that many architects do not think through the problem. The plans show a retaining wall, but they do not provide details for waterproofing behind that retaining wall. In areas where retaining walls are quite tall, a substantial amount of excavation, shoring, waterproofing, and backfill is required for a proper installation, and yet nothing appears on the plans.

Failure to install waterproofing can have serious consequences. First, to reiterate the point made above, the seepage of water through visible masonry or concrete walls sounds an alarm bell for owners and may drive them toward a construction defect lawsuit without any other problems in the building. Second, where wood frame structures have been built directly against retaining walls, moisture can and will migrate through the wall, into the framing, and into the interior of the structure. This can lead to an expensive repair process. In one notable case handled by the author, the recommendation for the repair of this condition for a large and elaborate tennis pavilion was the demolition of the entire structure, at a cost of more than $400,000. Third, the increased hydrostatic pressure against retaining walls caused by poor drainage can cause such walls to rotate out-of-plumb or to fail entirely.

Retaining walls that will have a finished surface, whether it is stucco or paint, or retaining walls that will directly contact a building structure must have an effective positive-side waterproofing. A positive-side waterproofing would include a perforated drain system; a waterproofing system, whether liquid applied or sheet membrane applied; adequate protection board; and, where feasible, a swale to divert water away from the area directly behind the retaining wall.

Figure 5-13 shows a large retaining wall that was only two months old at the time the photograph was taken. A large amount of visible efflorescence appears on this wall, and it is likely that the homeowner is quite unhappy.

Figure 5-13

Retaining wall without waterproofing

Here's how to avoid the defect:

- The plans must show a protocol for waterproofing masonry walls—preferably a positive-side waterproofing system.
- If the owner decides not to install a waterproofing system, obtain in writing from the owner his or her acknowledgment that he or she has been advised of the consequences of not installing a system and has elected not to proceed with the installation at no liability to the contractor.
- Obtain a specification from the architect or waterproofing consultant.
- Follow the specifications strictly, install the system carefully, test the system, and protect the work.
- Have the work inspected by the consultant or architect.

Waterproofing Balcony Decks

In the overall list of major factors that contribute to construction defect litigation, balcony decks are at the top. Many construction defect lawyers have a saying: "Every balcony deck leaks." There is, unfortunately, a great deal of truth to this.

Why do balcony decks leak? There is no single answer, as usually a combination of factors are involved.

First, as noted earlier, the framer often fails to slope the balcony deck framing properly. Either the framing improperly slopes toward the structure (instead of away from the structure) or the framing does not provide adequate slope toward the drains. When improper or inadequate slope is present, the deck is susceptible to more problems than usual, even if attention is paid to waterproofing. Second, the waterproofing of a deck requires a careful combination of many different building components, each of which has to be installed correctly and each of which must integrate with surrounding elements to perform properly.

Balcony deck waterproofing systems are just that: systems. They start with the proper preparation of the deck itself. The wood structure must be clean, smooth, and free from protrusions from nails, screws, and edges of plywood sheathing that could adversely affect the installation of the waterproofing membrane. The parapet walls must be blocked at the lower 12 inches so that the liquid applied membrane prep coat and *L* metal flashing has a secure base for attachment on the vertical leg. The flashing must be continuous at corners to avoid water intrusion, as shown in Figure 5-14.

After the deck surface is prepared, a prep coat of liquid-applied waterproofing material is installed around the edge of the deck at the transition from the horizontal deck material to the vertical parapet walls. This prep coat should cover the sill pan at doors that are adjacent to the deck.

After the prep coat is applied, the *L* metal that bridges the horizontal/vertical transition is embedded into the prep coat. To perform properly, this *L* metal flashing must be continuous—i.e., it must be soldered or otherwise installed so that it serves as a monolithic barrier to water. The *L* metal must lap vertically up the side of the walls at least 9 inches and must run horizontally onto the deck surface at least 6 inches. Builder paper must be carefully lapped over the top of the *L* metal.

Figure 5-15 shows a typical installation of a weep screed and *L*-base flashing.

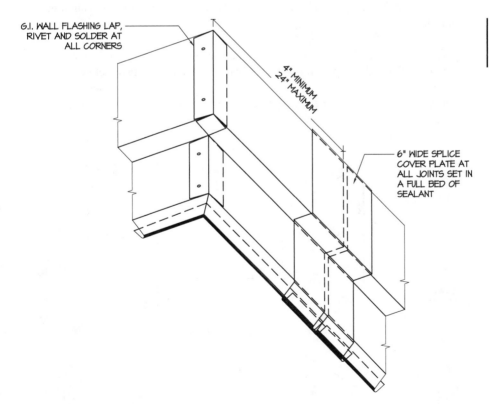

G.I. WALL FLASHING LAP, RIVET AND SOLDER AT ALL CORNERS

4" MINIMUM 24" MAXIMUM

6" WIDE SPLICE COVER PLATE AT ALL JOINTS SET IN A FULL BED OF SEALANT

Figure 5-14

Wall flashing properly installed

Figure 5-15

Base flashing at
weep screed

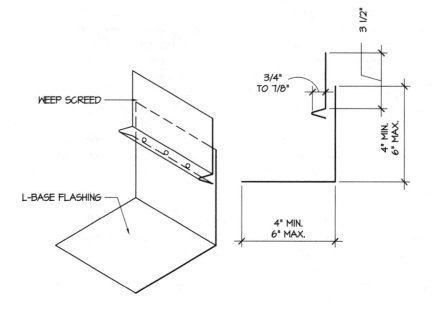

An alternative is to use a *diato* flashing, a sheet metal device that combines a weep screed with *L* metal, as depicted in Figure 5-16.

Figure 5-17 shows a failure to cover the *L* metal with builder paper, leaving a gap for water intrusion.

Figure 5-18 shows a deck to wall intersection where the *L* flashing was completely omitted, leaving a direct path for water intrusion into the building structure.

Figure 5-19 shows improper *L* metal installation.

Figure 5-16

Diato flashing

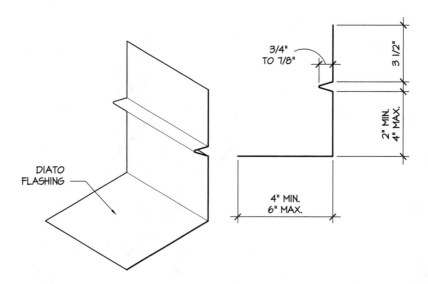

Figure 5-17
Builder paper is too short

Figure 5-18
Decks without *L* metal

Figure 5-19
Improper *L* metal

Figure 5-20 shows the damage that results from the lack of *L* metal flashing at a deck/wall intersection.

The installation and integration of the sill pan at the balcony door is critical to the performance of the system. To perform properly, the sill pan must be monolithic, and the vertical legs of the sill pan must be of sufficient height to repel water that might enter

Figure 5-20
Damaged deck

onto the surface of the sill pan. Problems in sill pans usually occur because the vertical legs at the ends of the pan are simply cut and bent, leaving a gap for water to enter. In many instances, the vertical legs are soldered to make a continuous seal, but the vertical legs are too short. Water enters the sill pan and runs over the top of the vertical legs and into the structure of the building.

Figure 5-21 shows a sliding door in a high-rise hotel that was installed without any sill pan at a balcony location. This condition permitted water to flow directly under the door threshold into the guest room.

Figure 5-22 shows an attempt to install a sill pan, but the installer has cut the vertical leg too short, leaving a path for water intrusion.

Figure 5-21
Damage from no sill pan installed

Figure 5-22

Water intrusion is inevitable

Figure 5-23 shows the repair of the condition shown in Figure 5-22. In these photographs, a new sill pan and adjacent *L* metal flashing has been installed under the door frame.

Even when sill pans are properly installed, they are still subject to problems usually caused by penetrations made after installation. Either the sill pan itself is penetrated by a nail or screw to adhere it to the framing or the sill pan is penetrated by a nail or screw when the door threshold is installed. Any penetration of a sill pan provides a path for water intrusion.

Typically, galvanized metal is used to form sill pans. It is my experience that galvanized metal will not withstand the test of time. I have seen repeated instances where galvanized metal sill pans have rusted through in only 5 years, resulting in significant water intrusion into the building. My recommendation is that leaded copper be used for sill pans. Leaded copper does present certain challenges, among them that the material is expensive (and sheet metal fabricators generally do not stock the material) and that care must be taken so that the leaded copper does not contact materials that could cause galvanic corrosion.

An alternative material, the pre-formed PVC sill pan, has recently come on the market. These sill pans are manufactured in three sections—two end sections and a middle section. The two end sections have vertical turn-ups that are pre-formed at the factory. To obtain the proper width, the middle section is set down on the framing and the two end sections are laid on top of the middle section at the appropriate distance. The assembly is then glued together with PVC cement, forming a monolithic assembly that resists corrosion and deterioration. One drawback to this system is the sill pan that is created by gluing the three pieces together does not have a built-in slope to drain toward the outside of the building. This requires the installation of shims under the sill pan to create positive drainage to the outside of the building.

Figure 5-23
New sill pan

Decks are drained either through area drains or scuppers, or using a combination of both, where the deck drains are used as the principal means of drainage and the scuppers are used as a secondary means of drainage in the event that the deck drains become clogged. Balcony deck waterproofing systems, no matter what type and no matter what quality, will fail if subjected to standing water for any length of time. Thus, for these systems to perform properly, water must be directed to the deck drain and the deck drain must function properly.

Most balcony deck drains are of a "double hub" construction. This type of drain has two means of drainage: one at the top surface of the deck, flush with the finished decking material, and the other at the level of the waterproof membrane. The bottom drain has a large metal flange that is bolted directly to the wood deck (in good quality installations, the flange would be placed on top of a prep coat of liquid applied membrane). The flange has a series of grooves that slope inward toward the drain pipe, terminating in a small drain hole. The purpose of the groove and drain system at the level of the flange is to provide a method to drain water that soaks through the finished surface of the deck and runs off the waterproof membrane toward the deck drain.

Figure 5-24 shows a specification for a double hub deck drain in a retrofit/repair situation.

Figure 5-25 shows a "low profile" deck drain. The large circular opening is the primary drain and will be connected to a pipe that will end at the finished elevation of the deck. The flower shaped flange surrounding the circular opening contains the secondary drains. If these secondary drains are clogged, water will pond under the finished deck surface and eventually leak into the structure below.

For both drains to function properly, the deck itself must slope toward the drain (so that any water that penetrates the finished decking surface material will flow across the waterproof membrane toward the deck drain), and the top surface of the deck must slope toward the primary drain inlet. In a proper installation, most of the water will flow

Figure 5-24

Double hub deck drain

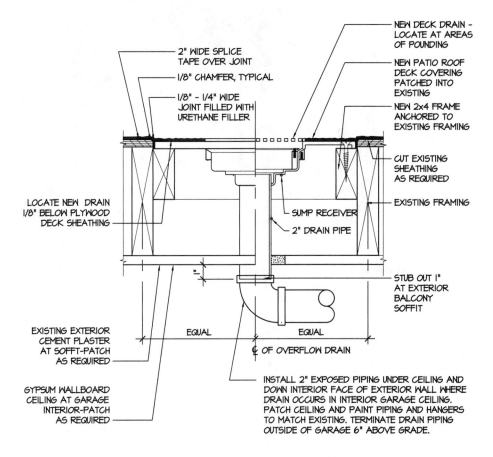

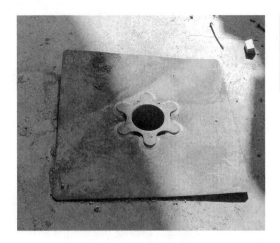

Figure 5-25
Low profile deck drain showing primary and secondary drain holes

toward the primary drain inlet at the surface of the drain, thereby minimizing the amount of water that saturates through the finished surface (i.e., through grout joints).

When decks and deck drains are not installed properly, the framing is not sloped toward the drain and the finished deck surface is not sloped properly. Water then ponds on the finished surface, creating a situation where the water is more likely to saturate through the finished surface. If the underlying deck is not sloped toward the drain, the water forms a lake under the finished surface and on top of the waterproof membrane. The water will remain in this condition until it dries out through evaporation or until it penetrates the waterproofing system (which is usually the case). A visible symptom of this condition is efflorescence on the surface of the deck, caused by water saturating the finished surface and then evaporating, leaving soluble salts on the surface.

Even in circumstances where the decks are properly constructed, drain holes in the bottom drain assembly can become blocked with waterproofing material, mortar, or grout. If this occurs, the water will drain toward the secondary drain but the water cannot escape into the drain assembly. To reiterate, water that is allowed to form a pond on the surface of the waterproof membrane and stand for any length of time will eventually cause the membrane to fail and water intrusion will result.

When scuppers are used as the primary drain system, the defect most often found is that the scuppers are set too high on the wall. It is often the case that the installer fails to flash around the sheet metal scupper, leaving a path for water to intrude into the balcony parapet wall. If the scupper is too high on the parapet wall, water stands on the balcony deck because it cannot drain out of the scuppers.

Figure 5-26 shows a scupper that was installed without flashing, leaving a substantial gap for water intrusion.

Figure 5-27 shows two scuppers that have been installed in a roof parapet wall. In this better application, the installer has flashed around the scupper using a peel-and-stick

Figure 5-26

Improper scupper
installation

membrane material. A better installation would have incorporated a sheet metal flashing around the scuppers.

Figure 5-28 shows a better application, involving a continuous flange soldered to the scupper assembly. In this configuration, the scupper assembly can be completely waterproofed where it meets the building surface by the use of a peel-and-stick membrane.

Related to all of these problems are two additional typical problem areas: the installation of balcony deck railings, and the waterproofing of balcony parapet walls.

Figure 5-27

Sealed scuppers

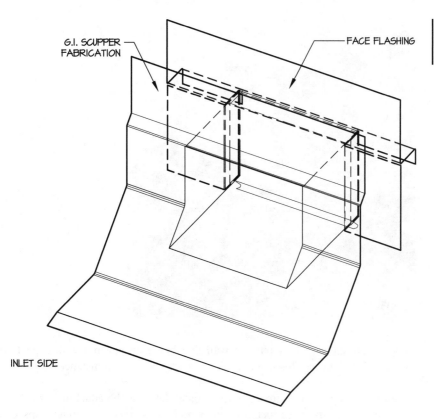

G.I. SCUPPER FABRICATION

FACE FLASHING

INLET SIDE

Figure 5-28

A better scupper application

Balcony deck railings, such as wrought iron railings, must be mechanically attached to the deck surface. Construction defects usually arise from the fact that these railings are often attached after the balcony deck is waterproofed. The mechanical attachment, usually in the form of lag screws, penetrates the waterproof membrane, creating a path for water intrusion.

Defects in balcony parapet walls almost always involve the improper use of builder paper to lap over the horizontal surface of the parapet wall, improper installation of metal coping caps, and failure to transition from the horizontal top of the parapet wall to the vertical wall of the structure adjacent. In the typical improper installation, the balcony parapet wall is framed from 2×4 or 2×6 lumber. Builder paper is wrapped over the top of the horizontal framing member that forms the top of the parapet wall. Lath and stucco are installed over the builder paper.

When it rains, water saturates the stucco on top of the parapet wall. Because the builder paper is horizontal, the water ponds on top of the paper, under the stucco and lath. Eventually, because builder paper is not meant to resist standing water (but only to shed water in vertical installations), the builder paper will rot. Once it rots, water will flow directly into the framing of the building. The solution to this problem is to cover the top of the parapet wall with a peel-and-stick type membrane, a liquid waterproofing membrane, or a metal coping cap.

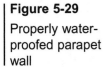

Figure 5-29

Properly water-proofed parapet wall

Figure 5-29 shows a roof parapet wall that has been covered with a peel-and-stick membrane at the top. This installation is proper for the circumstances.

When metal coping caps are used, the typical defect is the attachment of the metal to the structure or the improper joining of individual pieces. Coping caps cannot be attached by mechanical means from the top, because any rivets or screws on the top of the coping cap provide an immediate path for water intrusion (no matter how carefully they are "sealed" with various seam sealers, caulks, etc.).

Figure 5-30 shows two alternative methods of installation of coping caps (at roof level parapet walls).

Coping caps are subject to a substantial amount of thermal expansion and contraction (surface temperatures of sheet metal in the Western states can exceed 180°F). Unless the joints between the individual pieces of metal are fashioned to take thermal expansion into consideration, the joints will eventually break apart, leaving a gap and a path for water intrusion.

Figure 5-31 depicts the improper installation of sheet metal coping caps on a large custom home. These coping caps were placed on top of parapet walls in 8 foot sections, using a simple lap joint secured by sheet metal screws driven through the top of the sheet metal into the plywood below. In this installation, the builder also failed to install any water-proofing device—e.g., a peel-and-stick membrane—under the sheet metal coping caps. When water intruded through the joints of the coping caps, the water ran directly into the building structure, saturating the building from the roof level down three stories into the basement, where it caused significant mold growth. The photographs show substantial dry rot in the plywood, despite the fact that the home in question was only three years old.

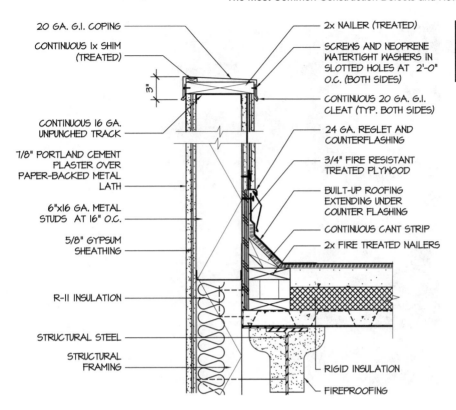

20 GA. G.I. COPING

CONTINUOUS 1x SHIM (TREATED)

CONTINUOUS 16 GA. UNPUNCHED TRACK

7/8" PORTLAND CEMENT PLASTER OVER PAPER-BACKED METAL LATH

6"x16 GA. METAL STUDS AT 16" O.C.

5/8" GYPSUM SHEATHING

R-11 INSULATION

STRUCTURAL STEEL

STRUCTURAL FRAMING

2x NAILER (TREATED)

SCREWS AND NEOPRENE WATERTIGHT WASHERS IN SLOTTED HOLES AT 2'-0" O.C. (BOTH SIDES)

CONTINUOUS 20 GA. G.I. CLEAT (TYP. BOTH SIDES)

24 GA. REGLET AND COUNTERFLASHING

3/4" FIRE RESISTANT TREATED PLYWOOD

BUILT-UP ROOFING EXTENDING UNDER COUNTER FLASHING

CONTINUOUS CANT STRIP

2x FIRE TREATED NAILERS

RIGID INSULATION

FIREPROOFING

Figure 5-30

Two methods of installing coping caps

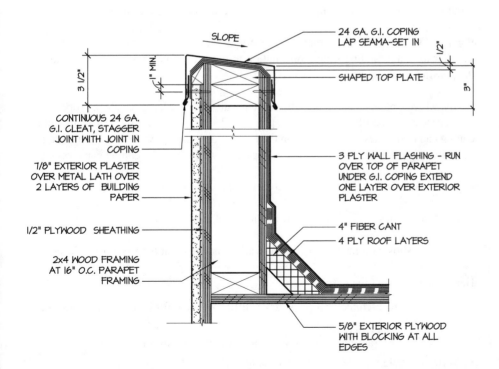

SLOPE

3 1/2"

1" MIN.

CONTINUOUS 24 GA. G.I. CLEAT, STAGGER JOINT WITH JOINT IN COPING

7/8" EXTERIOR PLASTER OVER METAL LATH OVER 2 LAYERS OF BUILDING PAPER

1/2" PLYWOOD SHEATHING

2x4 WOOD FRAMING AT 16" O.C. PARAPET FRAMING

24 GA. G.I. COPING LAP SEAMA-SET IN

1/2"

3"

SHAPED TOP PLATE

3 PLY WALL FLASHING - RUN OVER TOP OF PARAPET UNDER G.I. COPING EXTEND ONE LAYER OVER EXTERIOR PLASTER

4" FIBER CANT

4 PLY ROOF LAYERS

5/8" EXTERIOR PLYWOOD WITH BLOCKING AT ALL EDGES

Figure 5-31
Failed metal
coping caps

Figure 5-32 shows a similar failure in a commercial high rise building.

Figure 5-33 shows a sheet metal cover for a roof mounted hvac installation. The installer used a "standing seam" installation to produce a superior result and eliminate the possibility of failure at the seams.

Lastly, where a parapet wall meets the vertical wall of the building facade, a proper flashing is essential. Too often, a coping cap is installed on top of the parapet wall, and the coping cap terminates at the stucco on the face of the building, without a flashing to prevent water from intruding at the end of the parapet wall where the coping cap terminates.

Figure 5-34 shows a typical horizontal to vertical wall intersection.

Figure 5-35 shows a proper coping transition flashing.

Figure 5-36 shows horizontal to vertical wall transitions that have not been flashed.

Here's how to avoid the defect:

- Verify that the architectural plans have sufficient slope for balcony deck drains and scuppers.
- Verify that the framing slopes to drain at both deck drains and scuppers.
- Verify the height of deck drains and scuppers.

- Follow the instructions of the waterproofing consultant with regard to the installation of the balcony deck waterproofing system.
- Verify that the primary and secondary drain systems are functioning (not blocked with debris or mortar).
- Verify that balcony parapet walls are capped with appropriate waterproofing devices.
- Verify that flashings are installed at the termination of parapet wall coping caps.
- Verify that wrought iron or other railing materials are waterproofed at the point of attachment to parapet walls, other vertical walls, and at the point of attachment on the horizontal balcony deck surface.

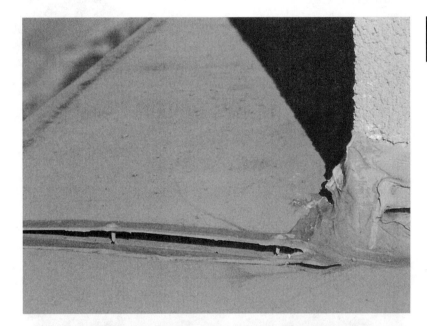

Figure 5-32
Photos of failed metal coping caps

Figure 5-32
(*Continued*)

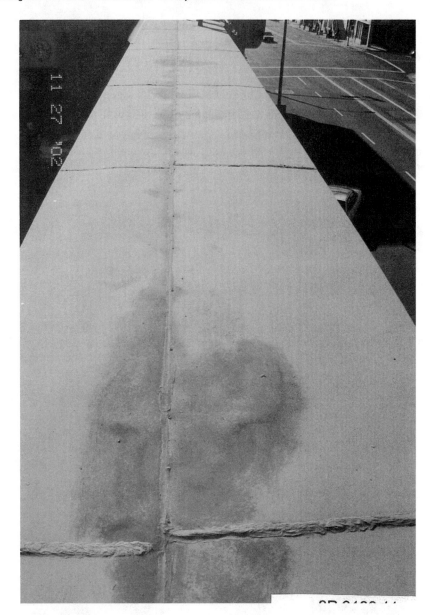

Waterproofing Architectural Plant-ons and Pre-cast

Many architects specify the use of architectural *plant-ons* (decorative elements such as raised bands that run horizontally around the building structure) or the use of concrete *pre-cast* elements around doors, windows, and at cornices. The attachment and the waterproofing of these elements is a critical means of avoiding construction defects.

Architectural plant-ons are usually made of framing lumber or Styrofoam. In either event, they are generally attached to the building prior to the installation of the builder

Figure 5-33
Standing seam roof

Figure 5-34
Typical horizontal to vertical wall intersection

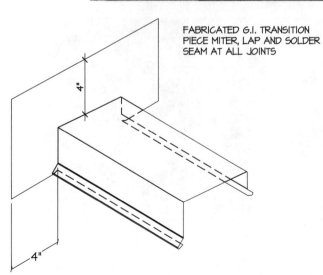

FABRICATED G.I. TRANSITION
PIECE MITER, LAP AND SOLDER
SEAM AT ALL JOINTS

4"

4"

Figure 5-35
Proper coping transition

Figure 5-36

Transitions without flashings

Figure 5-36
(*Continued*)

paper and lath. In improper installations, builder paper is wrapped over the top of the plant-on, so that the builder paper runs horizontally along the top of the plant-on. As noted with regard to balcony parapet walls, when water saturates the stucco on top of the plant-on and ponds on the horizontal surface of the builder paper, the builder paper will eventually rot, allowing water to intrude into the structure of the building. A proper installation uses a suitable flashing, such as a peel-and-stick membrane or sheet metal and the installation of a cant strip to provide positive drainage at the top of the plant-on.

The installation of concrete pre-cast architectural elements presents a different challenge. Pre-cast elements are heavy and they must be attached to the building structure in a secure manner. Often these elements are attached to the building structure after the builder paper and lath have been applied to the building. The various pins, straps, and other means of attachment are then installed, penetrating the builder paper and providing a path for water intrusion.

Many times, pre-cast elements are placed on top of masonry walls. Because the pre-cast concrete cap is porous, any water that penetrates the cap will ultimately penetrate the masonry wall underneath. If the wall is painted or covered with stucco, water will penetrate the masonry wall behind the paint and stucco and lift those materials off the surface unless a waterproofing membrane is installed under the pre-cast cap and on top of the masonry wall. Figure 5-37 shows a pre-cast concrete cap installed on a mortar bed on top of a masonry wall (that will eventually be covered with stucco). This installation will result in the failure of the stucco system.

Here's how to avoid the defect:

- Obtain details from the architect or owner's waterproofing consultant for any decorative elements to be attached to the building.
- Use appropriate flashing material to waterproof architectural plant-ons and pre-cast elements.
- Do not penetrate the building envelope when installing attachments for pre-cast elements or carefully seal around all such points of attachment.
- Strictly comply with the architect's recommendations and those of the owner's waterproofing consultant.
- Have the work inspected by the architect or waterproofing consultant.

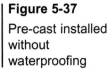

Figure 5-37
Pre-cast installed without waterproofing

Waterproofing Building Penetrations

Electrical conduit, dryer vents, exhaust vents, electrical junction boxes, and water lines can all penetrate the building envelope, and all of them present potential paths for water intrusion. Generally speaking, any penetration of the building envelope that occurs after the building has been wrapped with builder paper will create a problematic installation because it is difficult to integrate the pipe, vent, or electrical box with the builder paper.

The waterproofing of penetrations of the building are a continuing and substantial problem in the construction industry, but in most instances, they are unrecognized by contractors. The problem is twofold. First, no commercial products are available to waterproof easily around the many different building components that penetrate the building structure. Second, the nature of the components themselves makes them difficult to waterproof.

For example, even electrical junction boxes installed on the exterior of buildings, which are conventional and routine in construction, are in fact difficult to waterproof. To waterproof them correctly requires meticulous attention and a thorough familiarity with different types of waterproofing materials (not all caulking materials are compatible with plastic and metal J boxes and builder paper).

Figure 5-38 shows exterior J boxes that were not sealed correctly, leading to substantial water intrusion.

Figure 5-39 shows an exterior J box in new construction. The face of this J box has been installed flush with the surface of the surrounding shear panel. There is virtually no way to waterproof this installation.

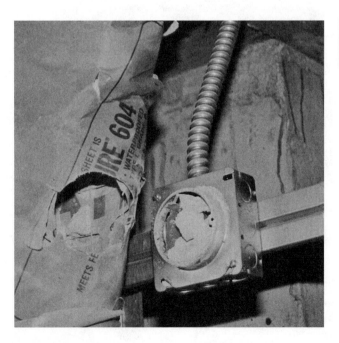

Figure 5-38
Incorrectly sealed J boxes

Figure 5-38
(*Continued*)

Figure 5-39
There is no way to waterproof this J box installation

Figure 5-40
Water will intrude around the edges of this J box

Directly related to this is the issue of retrofitting during remodeling of an existing structure. Figure 5-40 shows a common error: a J box that has been installed by chipping away stucco (with the idea that after the new J box has been installed, the stucco will be patched). Because the stucco is porous, water will intrude directly around the edges of the new J box, because the edges of the J box have not been sealed to the builder paper.

Similarly, it is difficult to make a mechanical connection between the outside of a pipe (for example, a water line for a hose bibb) and the building structure. The same could be said for PVC clean-out connections, dryer vents, exhaust vents, gas lines for outdoor barbeques, wood beams for decorative or structural support, or even openings for wires for cable television and telephone lines.

Figure 5-41 shows utility boxes mounted on balcony parapet walls on the exterior of a building. The builder simply screwed the boxes onto to the framing, and lapped roofing material around the box, leaving a direct path for water intrusion.

Figure 5-42 shows a somewhat similar condition—one that is very common. This photograph depicts an access door that was installed by fastening the door assembly to the framing and butting builder paper to the edges of the door without the use of any flashing or waterproofing material and then covering the builder paper with lath and stucco. When water saturates the stucco, it will run directly into the gap between the builder paper and the access door, and then into the building structure.

In concept, the approach to waterproofing all of these penetrations is the same. A waterproofing material must be installed around the outside of the component and that waterproofing material must be mechanically connected to the surface of the building before the builder paper is wrapped onto the building. To use the dryer vent as an example,

Figure 5-41

Improperly installed utility boxes

a peel-and-stick membrane could be attached to the outside of the vent and then applied to the surface of the building. To waterproof this kind of installation correctly, however, the peel-and-stick membrane would have to be installed carefully, and the vent and face of the building would have to be primed so that the membrane adheres tightly.

Because each type of building component that penetrates the building structure creates its own individual challenges for the contractor, the best approach would involve the use of a waterproofing consultant to devise a protocol for each individual component. Even

Figure 5-42
Water will run into the gap between the builder paper and the access door, into the structure

after the protocol has been developed, the installation of the waterproofing materials requires, again, meticulous attention to detail and the use of the proper materials.

Here's how to avoid the defect:

- Obtain a specification from the architect or waterproofing consultant with respect to all penetrations of the building envelope.
- Sequence and schedule the work so that penetrations of the exterior sheathing of the building are made prior to the application of the builder paper.
- Carefully follow all of the specifications and recommendations of the architect and waterproofing consultant.
- Have the work inspected by the architect or waterproofing consultant.

Waterproofing Around Windows and Doors

This is the number one cause of construction defects and construction defect litigation.

Defects in doors and windows always arise from the same problem: failure to install flashing and other devices properly to prevent water intrusion. After that it gets a little

Figure 5-43

Window with no flashing

more complicated. There are many reasons why windows and doors are not properly waterproofed. The most common are these:

- Failing to install any waterproofing material at all around the window or around the door jamb. Figure 5-43 shows a window that was installed in a commercial building without any waterproofing material. The wood framing of the building is clearly visible around the edges of the window assembly.
- Using the wrong material.
- Using the correct material but installing it improperly.

Figure 5-44 shows a window installation from a home constructed in the late 1960s. The window came from the factory with a sisal kraft flashing (that remains in relatively good condition). The installation, however, was improper. The installer applied builder paper to the edge of the window opening, and then installed the window with the attached sisal kraft flashing. This installation created a path for water intrusion from water entering under the sisal kraft and entering around the perimeter of the window opening, where the builder paper was not sealed to the window. The installer should have installed the window with its attached sisal kraft flashing first, then layered the builder paper over the window.

We must start from the position that every window and door opening must be waterproofed. Windows come from the factory with either a factory-applied flashing system or without any factory installed system. The integrity of the factory installed system must be respected. If no system is installed by the factory, one must be fabricated in the field.

Figure 5-44

Improper installation using sisal kraft flashing and builder paper

It is, unfortunately, all too common to see factory installed systems destroyed during the construction process. This is often the case in multi-family projects. Here's a typical scenario: A window system arrives from the factory with sisal kraft or another material used for flashing around the window. The windows are installed, and then the project sits for weeks, or even months, while other work progresses. At some point, the lathing subcontractor shows up to apply the stucco lath. By this time, the sisal kraft has been exposed to the sun for a few months, and it has curled up against the window frame. To apply the lath over the flashing, one worker has to hold the flashing down while another applies the builder paper and wire lath on top of it. Perhaps because not enough workers were available, they weren't trained, or they were just lazy, what often happens is that instead of having one worker hold the flashing down while the other applies the lath over it, the lathing contractor takes a razor knife, cuts off the flashing, and then butts the lath to the edge of the window frame. At this point, a leak is simply a matter of time, but the leak *will* happen, to a certainty, because there is a gap in the waterproofing system, leaving a clear path for water to intrude into the building.

Figure 5-45 shows a condition in which the sisal kraft flashing has been exposed to the weather for a considerable amount of time and has ripped and curled.

The simple step of holding down the sisal kraft flashing while applying the lath on top could avoid the problem, along with the hundreds of thousands of dollars that will likely be spent on the construction defect case that will surely follow.

Windows that do not have factory applied waterproofing systems require that a mechanical connection be made between the edge of the window and waterproofing material. In wood windows that do not have a nailing flange or nailing fin attached at the factory—i.e., a box frame window—a kerf or groove can be cut into the window edge, and the waterproofing material inserted into a bed of appropriate sealant. In metal windows made without a nailing flange or fin, the waterproofing material can be attached directly to the window after the appropriate primer has been used to insure that the waterproofing material will adhere. Note that the success of using peel-and-stick flashing material will depend on several things. First, the surface must be primed properly. If it is not, the flashing material will not adhere to the window, leaving space for water to intrude. Second, the correct type of peel-and-stick material must be used. For example, several

Figure 5-45

Sisal kraft flashing has ripped and curled after exposure

products have been used with excellent results, but only if the right product is used in the right place. As mentioned earlier, products used for below-grade installations should not be used above grade, or they will not withstand heat and will delaminate from the window frame, leaving a gap and a path for water intrusion. Many peel-and-stick membranes are available from various manufacturers, whose installation and use recommendations should always be thoroughly reviewed, understood, and followed.

I recommend that the inside of all window and door openings be covered with a peel-and-stick type membrane or sill pan flashing prior to the installation of the window or door. Once the window or door has been nailed or screwed into the window opening, additional flashing must be installed over the nail fin, if there is one, or the field-fabricated flashing must be attached to the window opening.

Figure 5-46 shows a good, but failed, attempt a waterproofing a window opening. In this photograph, the installer used moistop flashing under the window assembly. The finned window assembly was then installed. The failure of this installation will result from the fact that the installer nailed through the moistop flashing. Moistop flashing is not a self-sealing material, unlike some other materials.

In this part of the installation, the critical issue is the proper lapping, or shingling, of the material. A horizontal piece of membrane is installed under the sill of the window first, but only the top of this membrane is stuck onto the building surface. The bottom portion of the membrane is allowed to hang loose (so that builder paper can be slid underneath later). The two vertical pieces of waterproofing material that run along the

Figure 5-46

Moistop flashing is not a self-sealing material, and this installation will fail because nails penetrated the material

vertical sides of the window are installed next. A horizontal piece is installed next across the top, or head, of the window. The two vertical pieces must be lapped under the waterproofing material at the head of the window. The builder paper is then applied. At the sill of the window, the builder paper is slid underneath the peel-and-stick membrane, after which the membrane is stuck onto the builder paper.

Figure 5-47 shows the sequence of such an installation, using a moistop flashing material.

Figure 5-48 shows several examples of extremely thoughtful window installation techniques. The installer first covered the window opening with a peel-and-stick membrane. Next, the installer placed a copper sill pan at the bottom of the window opening. The copper sill pan is continuously soldered so that no gaps exist for water intrusion, and the sill pan incorporates a vertical leg on the interior side of the exterior wall (in essence, the window rests on and inside the sill pan). The gap between the window frame and the window opening was then filled with backer rod (not visible in the photographs) and another layer of peel-and-stick membrane was placed over the window fin and onto the surface of the building. The seams between each piece of peel-and-stick membrane were covered with seam sealer, and the edges of the peel-and-stick membrane were sealed to the building surface with seam sealer.

The installation shown in the sequence of photographs was part of the construction of a 30,000 square foot custom home, where the owner had expressed to the contractor the importance of avoiding any water intrusion into the residence. Although the installation

Figure 5-47

Proper installation sequence using moistop

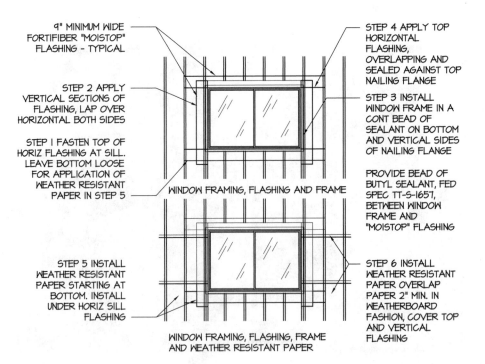

9" MINIMUM WIDE FORTIFIBER "MOISTOP" FLASHING - TYPICAL

STEP 2 APPLY VERTICAL SECTIONS OF FLASHING, LAP OVER HORIZONTAL BOTH SIDES

STEP 1 FASTEN TOP OF HORIZ FLASHING AT SILL. LEAVE BOTTOM LOOSE FOR APPLICATION OF WEATHER RESISTANT PAPER IN STEP 5

STEP 4 APPLY TOP HORIZONTAL FLASHING, OVERLAPPING AND SEALED AGAINST TOP NAILING FLANGE

STEP 3 INSTALL WINDOW FRAME IN A CONT BEAD OF SEALANT ON BOTTOM AND VERTICAL SIDES OF NAILING FLANGE

PROVIDE BEAD OF BUTYL SEALANT, FED SPEC TT-S-1657, BETWEEN WINDOW FRAME AND "MOISTOP" FLASHING

WINDOW FRAMING, FLASHING AND FRAME

STEP 5 INSTALL WEATHER RESISTANT PAPER STARTING AT BOTTOM. INSTALL UNDER HORIZ SILL FLASHING

STEP 6 INSTALL WEATHER RESISTANT PAPER OVERLAP PAPER 2" MIN. IN WEATHERBOARD FASHION, COVER TOP AND VERTICAL FLASHING

WINDOW FRAMING, FLASHING, FRAME AND WEATHER RESISTANT PAPER

Figure 5-48
Thoughtful window installation techniques

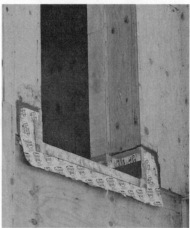

Figure 5-48
(*Continued*)

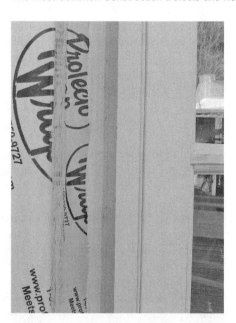

Figure 5-48
(*Continued*)

appears to be complex and expensive, in fact it would add little to the cost of installing windows in any structure (the largest single expense is the fabrication of the copper sill pans).

The installation shown in Figure 5-48 will be followed by pre-cast elements, and the copper sill pans incorporate a "step" for the installation of same, as shown in Figure 5-49. Thus, any water percolating through the concrete pre-cast will run off the copper sill pan and down the exterior of the building on top of the builder paper.

The failure to make a mechanical connection between the window assembly and the building structure, or to lap the waterproofing material properly, will result in water intrusion. Water intrusion will result in a lawsuit.

I believe that the better method of installation for windows that have box frames (i.e., no nailing fin) is, at a minimum, the installation of a peel-and-stick membrane on all four sides of the window opening, after the opening has been primed properly, followed by the installation of a sill pan as shown in the preceding figures. An even better installation would include a continuous pan around all four sides of the window (or the fabrication of such a device using peel-and-stick membrane—a laborious and time-consuming process). If a sill pan is used at the bottom of the window opening, the pan must be slightly canted toward the outside by the use of plastic shims. Any water that finds its way into the sill pan will thus be directed back onto the face of the building and run down the builder paper as intended.

In this regard, a somewhat unusual, but thoughtful, approach was taken with respect to the door installation in the large custom home shown in the preceding figures. At the

Figure 5-49

Water will run off the copper sill pan and down the exterior of the building, as it should

Figure 5-50
The beveled threshold installation

exterior door thresholds, the installer placed three layers of peel-and-stick membrane, starting from the innermost edge of the door and working outward. In other words, on the innermost portion of the door threshold, three layers of peel-and-stick membrane were followed by two layers and then one layer. In effect, the installer created a bevel or slope leading outward by this method. Figure 5-50 shows this installation.

The bottom of the door thresholds (custom fabricated) were cut at a 6 degree bevel to fit exactly flush with the beveled peel-and-stick threshold. Figure 5-51 shows the door threshold in section.

In this manner, the installer has insured that any water that comes into contact with the door threshold assembly will be directed back out to the exterior of the house and cannot intrude into the interior.

Figure 5-51
The door threshold in section

In addition, code requirements for clearance between deck surfaces and door thresholds should be carefully observed to avoid the flow of water from a deck surface into or under the door threshold. Most codes require a 3 inch clearance between the top of the deck and the bottom of the threshold. Oftentimes this does not leave enough room for the balcony deck to be sloped to drain away from the structure. This is an important plan detail, but it is frequently omitted from plans or simply drawn improperly. The plan details in these areas should be carefully reviewed and the contractor should confirm that sufficient height at the threshold has been provided.

In Figure 5-52, we see a variation on this theme. The photographs shows the threshold of an exterior French door leading to an exterior concrete patio area. Investigation revealed

Figure 5-52

Improperly applied flashing terminated at grade, so that wood framing was sitting in dirt

that the builder constructed the adjacent swimming pool before the main house structure was built. The concrete flatwork was to be at the same height as the pool coping. However, the pool and pool coping were 4 inches too high. After the house was constructed, the builder started to install the concrete flatwork and realized that it would cover the door threshold. The builder installed a heavy gauge galvanized flashing under the door threshold in an effort to prevent water from reaching the door threshold assembly.

Unfortunately, the builder did several things wrong. First, the galvanized flashing was not a continuous soldered piece of metal. The builder installed several pieces of flashing material, using overlapping joints, and nailed through the joints. Second, the builder simply lapped the flashing vertically down the face of the assembly, but terminated the flashing at grade. The result, as shown in the figure, was that the wood framing was sitting directly in dirt, and buried below the concrete flatwork, which collected water.

The installation of windows often leads to failure of the waterproofing system. For example, solid vinyl windows have a substantial capacity to expand and contract. The nailing fins on solid vinyl windows have pre-formed slots instead of holes for nailing. The design concept is that the nail fin will have the ability to move slightly due to the slot configuration. Many times the installer will simply nail through the plastic nail fin, or will nail too close to the corners of the windows. When this happens, the thermal expansion of the window causes the window to move and the corners to crack.

Related to this problem is the issue of broken corners. Although the manufacturers of solid vinyl windows will state that the corners are quite strong, if windows are mishandled on the site prior to installation or jammed into window openings that are not square, the torque applied to the window frame can break the window at the corner, leading to water leaks.

Here's how to avoid the defect:

- Obtain specific details from the architect or owner's waterproofing consultant for the waterproofing of window and door openings.
- If not inconsistent with recommendations of the architect or waterproofing consultant, flash the inside of all window and door openings and install sill pans or four way pans.
- Verify that a flashing system is present for all window and door frames and that the integrity of the flashing system is not harmed.
- Carefully install the windows and doors to avoid torqueing the assemblies and follow all manufacturer's recommendations.
- Verify that the shingling or lapping of the flashing material is proper and that it integrates with the builder paper.
- Take steps to maintain the integrity of sill pans to avoid penetration of sill pans.

Roofing and Skylight Waterproofing Issues

Surprisingly, most construction defect water intrusion cases do not begin with roof leaks; they begin with leaks from windows, doors, and balconies. When roof issues arise, they nearly always have more to do with roofing details and with the installation of skylights.

Roofing Details

In built-up roofs, here are the usual suspects:

- Failure to attach the base sheet with the manufacturer's recommended spacing of roofing nails.
- Installation of interplys with asphalt that is too hot or too cold.
- Installation of base sheet and subsequent layers of roofing material on plywood or OSB that is too wet.
- Failure to install cant strips at horizontal/vertical intersections. Figure 5-53 shows a roof to wall intersection with the use of a cant strip and reglet flashing. Note that the plan calls for a minimum 7 inch distance between the bottom of the reglet flashing and the roof to avoid water intrusion through the bottom of the reglet flashing.

In other roofs, these are usual suspects:

- In slate or mission tile, or other tile roofs, failure to wrap the ridge board with builder paper.
- In slate, mission tile, or other tile roofs, failure to have the required overlap between courses of tile.

Figure 5-53

Flashing at wall

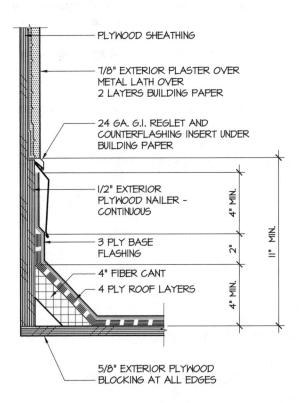

PLYWOOD SHEATHING

7/8" EXTERIOR PLASTER OVER
METAL LATH OVER
2 LAYERS BUILDING PAPER

24 GA. G.I. REGLET AND
COUNTERFLASHING INSERT UNDER
BUILDING PAPER

1/2" EXTERIOR
PLYWOOD NAILER –
CONTINUOUS

3 PLY BASE
FLASHING

4" FIBER CANT

4 PLY ROOF LAYERS

4" MIN.

2"

4" MIN.

11" MIN.

5/8" EXTERIOR PLYWOOD
BLOCKING AT ALL EDGES

- In slate, mission tile, or other tile roofs, failure to secure the tiles with a mechanical attachment.
- In all types of roofing, failure to install step flashing properly at a roof/wall intersection.
- In all types of roofing, failure to install a saddle flashing around chimneys.
- In all types of roofing, failure to slope the roof adequately to avoid ponding water.
- In all types of roofing, failure to install proper metal flashing at all roof "valleys."
- In all types of roofing, improper flashing of penetrations of the roof for vents, pipes, and hvac refrigerant lines.

Figures 5-54 and 5-55 show various failures to seal around roof penetrations properly. Figure 5-54 shows a pipe penetration and a roof jack. The roof jack was the proper approach, but the builder elected to install the roof jack on top of the finished roof surface, instead of under the surface, which is the proper location for a roof jack. Note also in this photograph that the gap between the pipe and roof jack is not sealed, creating a path for water intrusion into the building.

Figure 5-55 shows similar conditions. In these photographs, electrical conduits have penetrated the roof and have not been sealed to prevent water intrusion.

Figure 5-56 depicts proper installation procedures for pipes and plumbing vents that penetrate the roof assembly.

Skylight Details
The typical construction defect from skylight installations is that curb-mounted skylights are installed on curbs that are too low. Skylights that are mounted on curbs that are too low are prone to water intrusion from water ponding on the roof around the skylight openings.

Figure 5-54
Improperly installed roof jack

Figure 5-55

Electrical conduits penetrating the roof have not been sealed properly

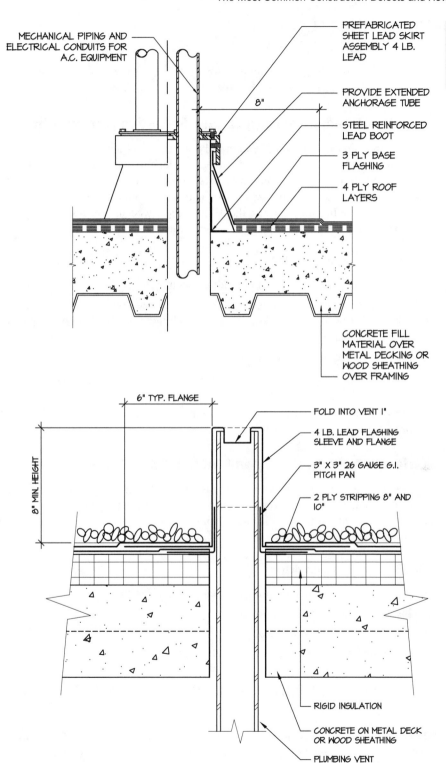

MECHANICAL PIPING AND ELECTRICAL CONDUITS FOR A.C. EQUIPMENT

PREFABRICATED SHEET LEAD SKIRT ASSEMBLY 4 LB. LEAD

8"

PROVIDE EXTENDED ANCHORAGE TUBE

STEEL REINFORCED LEAD BOOT

3 PLY BASE FLASHING

4 PLY ROOF LAYERS

CONCRETE FILL MATERIAL OVER METAL DECKING OR WOOD SHEATHING OVER FRAMING

6" TYP. FLANGE

8" MIN. HEIGHT

FOLD INTO VENT 1"

4 LB. LEAD FLASHING SLEEVE AND FLANGE

3" X 3" 26 GAUGE G.I. PITCH PAN

2 PLY STRIPPING 8" AND 10"

RIGID INSULATION

CONCRETE ON METAL DECK OR WOOD SHEATHING

PLUMBING VENT

Figure 5-56

Pipe penetration, and flashing at a plumbing vent

DEFECTS IN DRYWALL INSTALLATION

Two types of drywall defects are often noted: defects related to fire-resistive construction, and defects related to attaching drywall.

Drywall Defects Relating to Fire-Resistive Construction

Fire-resistive construction requires not only attention to framing, but also attention to the installation of drywall to achieve a rated envelope to delay the spread of fire within the building structure. The common defects are failure to install two layers of drywall at party walls in multi-family construction and failure to maintain a one-hour envelope in garage spaces. In garage spaces, the defect typically occurs because openings are cut into the envelope, and those openings are either not sealed or inadequately sealed.

Here's how to avoid the defect:

- Install the required number of drywall layers at party walls.
- In garages and other enclosed spaces where a one-hour fire separation must be achieved, drywall joints and openings through the drywall must be completely and carefully sealed.
- Provide the required rated door assembly at rated walls.
- If air conditioning ducts penetrate a rated wall, an approved fire rated damper must be used.
- Plastic or PVC pipe penetrations are not permitted in fire-rated assemblies; copper or cast iron pipe only must be used.

Defects Relating to Drywall Installation

Building codes require that drywall be fastened to the framing of the structure with screws or nails of a certain distance apart. A typical defect claim alleges that the fasteners are spaced too far apart. Other common defect claims for drywall include nails pops and cracks along corner beads.

Here's how to avoid the defect:

- Follow building code requirements for spacing of fasteners.
- Install drywall with screws not nails.
- Adequately secure corner beads.
- Verify that corner beads are square and plumb.

DEFECTS RELATING TO EXTERIOR PLASTER SYSTEMS (STUCCO)

Defects relating to stucco installation are among the most common of all construction defects. More importantly, from a liability perspective, defects relating to exterior plastering systems are perhaps among the simplest to prevent but the most important to avoid for a very simple reason: homeowners easily see stucco cracks, and they assume if the stucco is cracked that a defect exists. Once the owner becomes concerned that a single defect exists, the owner will generally begin to believe that other defects exist, and a lawsuit will not be far behind. In addition, the repair of stucco defects is extremely costly.

Fortunately, exterior plaster defects are simple to avoid. They fall into seven major categories:

- Failure to wrap the structure properly in builder paper
- Failure to attach the lath properly to the building
- Failure to mix the cement/sand in the proper ratio
- Failure to attach the plaster mechanically to the lath—i.e., to "key" the plaster into the lath
- Failure to keep the scratch coat and brown coat damp for the code-required 48 hour period and to patch cracks before applying the next coat of plaster
- Failure to install or correctly install weep screeds
- Failure to weatherproof penetrations of the building envelope

An understanding of the issue of wrapping the building requires an understanding of the concept behind waterproofing an exterior stucco system, something that appears to be poorly understood in the community of construction professionals. It is worth repeating yet again, stucco systems are designed with the idea that the stucco itself is a porous product that will not resist water for any appreciable amount of time. It is assumed that water will eventually penetrate through the stucco to whatever material has been installed under the stucco system. Thus, the water-resistant capacity of the building does not depend on the stucco. It depends on what is *under* the stucco.

The waterproofing system under stucco relies on the concept of *weatherboarding*. The concept is similar to shingles on a roof, although in the case of the material under stucco, shingles are not used. A heavy paper, saturated with asphalt, called "builder paper" is used instead. Two layers of the builder paper are stapled or nailed to the outside of the building. The layers of paper are shingled, or lapped, much like shingles on a roof, so that any water that comes into contact with the builder paper will flow downwards and eventually drip out of the bottom of the weep screed placed at the bottom of the exterior stucco wall. In essence, this application creates a "vertical roof."

In much the same way that water would leak through a roof if the shingles were not lapped properly, water will leak through the walls of a structure if the builder paper is not lapped properly. To provide the requisite water-resistant system, all of the layers of builder paper must be properly lapped, the areas where there are vertical seams must be overlayed sufficiently, and the paper must be installed so that there are no rips, tears, or other entry points for water. Any "reverse lapping" of builder paper, or any significant tear or rip in the builder paper, will immediately lead to water intrusion in the building.

Figure 5-57 shows poorly lapped and torn builder paper.

Figure 5-57

Poor lapping and tears

The builder paper must also be integrated with the waterproofing devices around door and window openings and, as noted, other penetrations of the building, such as exterior J boxes for outside electrical fixtures, hose bibbs, and vents. Builder paper that is cut too short around penetrations will lead to water intrusion.

Figure 5-58 shows paper that was cut too short at a deck installation, leaving a path for water intrusion.

Figure 5-58

Paper was cut too short at a deck installation

The typical problems encountered in defective installation include reverse lapping of builder paper, failure to use two layers as required by code, failure to use the proper type of builder paper (Type D or 60 Minute paper), tears and rips in the paper, and failure to integrate the paper to other waterproofing materials.

With regard to the application of the lath, the UBC requires that lath be attached to the framing members of the building (not the sheathing) with a minimum of 7/8 inch embedment into the framing member. For metal studs, the sheet metal screw must be attached to the stud with a minimum of three turns of the screw.

All too often, because the lather does not mark the lath with the location of the studs (the "framing members")—i.e., the lather does not "chalk line" the exterior of the building to mark the location of the studs—the lath is attached only to the sheathing of the building. This installation will result in cracking to the finished exterior plaster system because a nail or staple driven into sheathing (and not driven 7/8 inch into the studs) does not have the ability to secure the lath strongly enough to the building. As a result, the lath will begin to move away from the building and stucco cracks will develop (recall that a standard three-coat stucco system weighs about 14 pounds per square foot: even a small, 10×20 foot section of wall thus weighs well over a ton).

A failure to attach the lath adequately to the studs will usually manifest itself in a characteristic crack pattern on the exterior of the building. This pattern will show regular vertical and horizontal cracks at regular distances. The cracks usually appear at the horizontal and vertical edges of the pieces of lath, particularly when K-Lath or equivalent is used. K-Lath is a product that combines builder paper and wire lath in one unit. The panels come in 8 or 10 foot lengths. When K-Lath is not properly attached to the structure, it is common to see vertical cracks every 8 or 10 feet and horizontal cracks every 4 feet, at the edges of the panels.

A related defect involves the mechanical attachment of the scratch coat to the lath. To hold the plaster securely to the lath, the scratch coat must be mechanically attached, or "keyed" into the lath. To achieve this mechanical attachment, the lath must be held away from the builder paper so that the plaster can be forced behind the lath during the application of the scratch coat. If the scratch coat is not forced behind the lath, and merely attached to the top of the wire lath, the plaster system may pull away from the lath or crack.

Figure 5-59 shows poorly embedded stucco.

The stucco mix is also critical to a defect-free installation. A standard three-coat system should be mixed in a ratio of about one part cement to three to four parts sand. Too much sand weakens the plaster to the point that it will break and flake off the building. Too little sand makes the plaster brittle, leading to breakage of the plaster system.

Hydration of the plaster is an essential element of a proper installation. Recall that in a three-coat system, a scratch coat is applied first, then a brown coat, and finally a finish

Figure 5-59

Poorly embedded
stucco

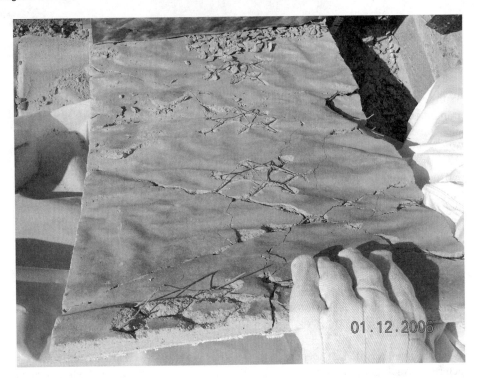

coat. The UBC requires that the scratch coat be kept moist for 48 hours before application of the brown coat. Likewise, the brown coat is to be kept moist for 48 hours before application of the finish coat. Often, particularly in the hot and dry Western states, the stucco subcontractor will not have a worker return to the job site to moisten the scratch or brown coat. This leads to a failure to hydrate the system and will create plaster that is weak, brittle, and prone to cracking. As a result, and particularly with respect to the brown coat, cracks will inevitably develop. Cracks in the brown coat, unless repaired, will "telegraph" through the finish coat.

It is my experience that the 48-hour cure time for the scratch and brown coats is substantially too short for a defect-free stucco installation. Cracks will develop in both the scratch and brown coats if they are allowed to cure for a longer period of time. Those cracks should be allowed to develop, and they should then be repaired to avoid the danger of telegraphing through the final finish coat. I recommend to builders that both the scratch and brown coats be allowed to cure for two weeks each. For production housing, this is difficult, but for custom housing it is a minimum to ensure a superior product.

The issue of weep screeds is a frequent cause of construction defect litigation. The weep screed, a metal piece with holes, serves two functions: it provides a hard, level, and flat edge, or screed, for the plasterer so that the edge of the plaster is straight and smooth; and it provides a means for water to drip, or weep, out of the bottom of the exterior wall onto a hard surface or the ground.

Figure 5-60
Weep screed

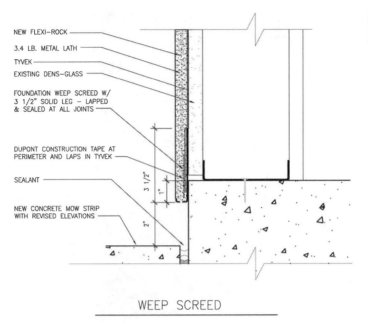

NEW FLEXI–ROCK

3.4 LB. METAL LATH

TYVEK

EXISTING DENS–GLASS

FOUNDATION WEEP SCREED W/
3 1/2" SOLID LEG – LAPPED
& SEALED AT ALL JOINTS

DUPONT CONSTRUCTION TAPE AT
PERIMETER AND LAPS IN TYVEK

SEALANT

NEW CONCRETE MOW STRIP
WITH REVISED ELEVATIONS

3 1/2"
1"
2"

WEEP SCREED

Figure 5-60 shows a weep screed in section.

The UBC requires that the weep screed be placed at least 2 inches above hardscape and 6 inches above softscape (dirt or landscaped materials). It is extremely common (too common given the simplicity of the code requirement) to find weep screeds buried in either dirt or concrete flatwork. Figure 5-61 shows a weep screed buried in concrete flatwork in a commercial building (this photograph also shows a common problem with the location of the rebar in flatwork—note that the builder has failed to install the rebar in the middle of the concrete—the rebar was placed on grade and is not reinforcing the concrete).

Figure 5-62 shows a weep screed buried in pavers.

Figure 5-63 shows an exterior stucco wall without a weep screed. The stucco has deteriorated to the point that it is flaking off the building due to the build-up of moisture inside the stucco.

Sometimes the weep screed is not buried but it is nonetheless too low to the ground. In Figure 5-64, a weep screed on a commercial building was installed only an inch or so above a concrete swale. Investigation of this condition revealed that the quantity of water in the swale was such that the water rose above the level of the weep screed, resulting in water intrusion into the building.

It is vital that the stucco contractor provide a path for water to weep from the building through the weep screed to avoid the build-up of water in the wall itself. If water is allowed to remain in the exterior plaster and not allowed to weep out of the wall through

Figure 5-61

Weep screed buried in concrete flatwork

Figure 5-62
Weep screed
buried in pavers

Figure 5-63
Deteriorating
stucco

Figure 5-64

Weep screed installed above a swale allowed water intrusion

the holes in the weep screed, water may penetrate the weatherboarding of the building and saturate the framing members and ultimately the drywall and other finished surfaces.

Stucco systems often incorporate decorative elements called *reglets* or *reveals*. These installations are a prime source for water intrusion. The reveal is generally made from extruded aluminum and consists of a *U*-shaped channel with a flange at the bottom of each side of the *U*. The flange is mechanically attached (i.e., screwed or nailed on) to the building structure. Usually, these channels, or reveals, are installed on top of the builder paper.

The difficulty with these installations is that builder paper must be perfectly intact below the installation. If there is any rip, tear, or opening in the builder paper, water will infiltrate through the seams in the channels (where the ends of the individual pieces are butted together) and will intrude into the framing of the building.

Figure 5-65 illustrates a defective installation of stucco reveals. The photograph shows the junction of the horizontal and vertical reveals being removed from the building during destructive testing.

Figure 5-66 shows the *butt joint* or *splice* in the horizontal reveal, leaving a path for water intrusion.

Figure 5-67 shows the back side of the splice. You can also see an example of poor stucco embedment in the lath.

Figure 5-65
Junction of reveals

Figure 5-68 shows a tear in the builder paper below the splice. This tear created a direct path for water intrusion into the building.

Figure 5-69 shows serious rust resulting from the water intrusion.

In many instances, decorative elements are installed as part of the exterior plaster system. In Figure 5-70, a large decorative pre-cast baseboard was installed at grade on the exterior of a home. However, no provision was made for water to weep from the wall above, and the water trapped in the wall will eventually either leak into the house or destroy the pre-cast decorative element.

Figure 5-66

The splice leaves a path for water intrusion

Figure 5-67

Back side of the splice

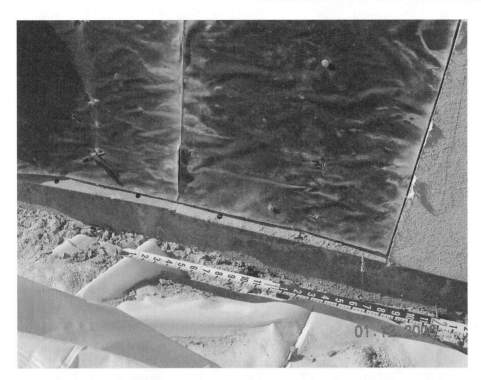

Figure 5-68
Torn builder paper

Figure 5-69
Rust resulting
from water
intrusion

Figure 5-70

Improperly
installed pre-cast

Avoiding construction defect claims on exterior plaster that has smooth, steel troweled, finish has it own challenges. Clients sometimes want this finish, but what they do not understand is that by nature a smooth, steel troweled finish is likely to show mottled colors and burn marks. The discoloration or mottling of stucco in a steel troweled finish is caused by the heat created by the friction between the trowel and the stucco as the trowel is drawn across the surface of the stucco during the finishing process. In addition, cracks in this type of finish are almost inevitable, though it is possible to minimize cracks in smooth stucco by the use of fiberglass mesh in the brown coat and various admixtures, but such installation techniques and materials will double or triple the cost of the work.

To avoid a claim for stucco cracks in steel-troweled finish systems, put the following in writing and have the client sign off:

> Client acknowledges and agrees that Client has been informed that steel troweled smooth stucco finishes will not produce a uniform color or finish and that they will show cracks. Contractor recommends that if Client desires a uniform color in the finish, that the exterior plaster be painted with an appropriate elastomeric paint system. Application of an elastomeric paint system will change the appearance of the finished surface and will more closely resemble a painted not plastered surface. Elastomeric paint systems, unless perfectly applied, and even if perfectly applied, will generally not prevent water intrusion into the exterior plaster. Large cracks will not be completely covered by elastomeric paint systems.

A relatively new product that can be used to minimize cracks in steel troweled systems is Isenwall, a three-coat application whose curing compounds are acrylic based and cure rapidly. All three coats can be applied in a matter of days, and the system is extremely resistant to cracking. Another proprietary product is Flexirock, which uses special additives to the stucco mix to minimize cracks.

A plaster specialist can specify any number of specialized formulae for exterior plaster for a given condition, but owners need to be aware that most stucco subcontractors are accustomed to commercially available products and that the price differential between a standard mix and a custom mix can be enormous to minimize cracks.

Here's how to avoid the defect:

- Verify that Type D or 60 Minute Paper is being used. Lap all builder paper properly to provide a shingled or weatherboarded exterior installation, and integrate builder paper with other waterproofing materials.
- On buildings with exterior plywood sheathing, the location of the studs should be clearly marked for the lathing contractor.
- The lathing contractor should be specifically instructed (and this instruction should be contained in the contract with the lather and in the plans, both of which should be signed by the lather) that the lath is to be attached to the framing members and not the sheathing.
- The general contractor should verify the size of the nails or staples being used to insure that the nails are embedded at least 7/8 inch into the framing members (recall that if the building is sheathed with plywood, the length of the fastener will have to be sufficient to go through the sheathing and into the framing).
- The general contractor should specify that the scratch coat is to be keyed into the lath, and the general contractor should verify that there is sufficient space between the wire lath and the builder paper to accomplish this mechanical attachment.
- The stucco contractor should be specifically instructed (and this instruction should be contained in the contract with the stucco contractor and the plans, both of which should be signed by the stucco contractor) that the scratch and brown coats are to be kept damp for at least 48 hours.
- If possible, the brown coat should be left on the building for at least 21 days before application of the finish coat.
- The brown coat should be carefully inspected by the general contractor to determine whether it is cracked. All cracks should be properly repaired before application of the finish coat.
- The general contractor should photograph the condition of the brown coat before application of the finish coat.

MISCELLANEOUS DEFECTS

Shower and Tub Enclosure Defects

A common and easily avoidable construction defect involves the construction of shower and tub enclosures. For decades, shower and tub enclosures were constructed using greenboard to cover the framing, followed by an application of a finished material such as tile or marble. The fundamental problem with this installation is that nearly all finished materials leak water through grout joints. When the water penetrates the grout joint, unless the surface below is waterproof, damage will result. Because greenboard is faced with paper made from wood pulp, mold grows readily on the greenboard surface.

There are various approaches to waterproofing shower and tub enclosures. One technique uses the following sequence:

1. Install cement backer board over framing.
2. Seal all seams with fiberglass mesh tape and mortar.
3. Apply a peel-and-stick membrane up at least 6 feet on the vertical walls and overlapped onto the hot mopped shower pan.
4. Install lath and mortar bed.
5. Install finished surface.

An alternative approach would be to use a liquid applied membrane (which must be self-sealing) over the cement backer board, or to use one of the proprietary PVC membrane systems that are chemically welded together into a monolithic sheet. All of these systems recognize the underlying fundamental issue of water intruding through the finished surface and the need to provide a waterproof membrane under the finished surface.

Defective Plans

With apologies to the many competent and careful architects, the fact is that architectural plans themselves are often the cause of construction defects. This occurs for a variety of reasons, including the following:

- The failure of the architect to coordinate the plans. *Coordinate* in this sense means several things:
  - Failure to perform a light-table analysis to ensure that the plans for the upper stories of a project coordinate with those for the lower story, so that pipes, columns, walls, and the like can be constructed without interference.
  - Failure to take into consideration the need for space for mechanical and utility installations. All too often an architect fails to leave room for ductwork, waste lines, or supply lines.
- Failure to allow clearance for the installation of recessed light fixtures, particularly when the installation of batt insulation is a necessary component. Where this occurs, it often becomes necessary to purchase expensive insulated fixtures.
- Failure to specify means of attaching pre-cast elements to the exterior of the building, leading to water intrusion through improper application techniques.
- Failure to provide sufficient slope to drain on balconies and terraces.
- Failure to provide sufficient height at exterior door thresholds.
- Failure to allow for the thickness of drywall at finish details.

Defects in Hardwood Flooring

Defects involving the installation of hardwood flooring usually involve the following:

- Failure to acclimate the hardwood flooring inside the building prior to installation.

- Failure to leave sufficient space around the boundary of the hardwood flooring installation to allow for expansion and contraction of the material.

- Failure to install a vapor barrier under the hardwood flooring in circumstances where damp conditions may damage the flooring.

Failure to acclimate the hardwood flooring material prior to installation will result in unacceptable gaps in the hardwood floor after installation, as the wood contracts through moisture loss. In addition, if too much moisture remains in the wood prior to installation, the individual planks will "crown or cup " because the moisture swells the bottom side of the plank but has no means to escape through the top which has been sealed.

Failure to allow sufficient boundary space around the perimeter of the hardwood floor installation will result in buckling of the individual pieces of hardwood floor. As the floor expands and contracts through weather cycles and through heating and cooling, it must have room to expand both longitudinally and laterally. Without sufficient space, the individual pieces of hardwood floor must move upward, since this is the only direction available.

In circumstances where moisture or vapor drive is present, the vapor barrier is essential to protect the material from excessive moisture. Excessive moisture will cause the floor to "cup" or buckle at the junction of individual pieces of flooring.

Here's how to avoid the defect:

- Acclimate the hardwood flooring material according to the manufacturer's recommendations and specifications and in accordance with the standards of the national trade organizations. Flooring must be acclimated in the environment in which it is to be used, meaning that it will require acclimatization in an air conditioned space in the summer and heated space in the winter if those are the conditions in which the flooring will eventually be used over the long term.
- Allow the code-required and manufacturer specified boundary space around the perimeter of the floor.
- Carefully install a vapor barrier (without perforations, rips, tears, holes, or failures to overlap the individual pieces of barrier adequately) in instances where vapor drive is present.

Acoustical Issues

Acoustical issues are a common cause of construction defect complaints, in both detached single family residences and in multi-family construction, such as attached townhomes and condominiums. Broadly speaking, acoustical concerns can be divided into the following categories:

- Through-wall noise, or noise that is transmitted from room-to-room through walls
- Floor-to-floor noise
- Plumbing noise
- Noise from mechanical systems, meaning hvac systems
- Outdoor noise through windows

By way of introduction, the UBC contains requirements to limit the amount of sound that is transmitted from floor-to-floor and from room-to-room. These sound transmission limitations are expressed numerically as an STC (Sound Transmission Classification) number. An STC of 50 represents a 50 percent reduction in the level of sound that would otherwise be transmitted from room-to-room or floor-to-floor but for the sound attenuation device.

In the "real world," architects, contractors, and owners use other, less technical means of determining what is an excessive amount of sound. In terms of floor-to-floor sound transmission, owners generally find the sound of footfalls to be unacceptable. In this regard, the amount of noise that an owner will tolerate seems to be directly proportional to the cost of the home or condominium. The more expensive the project, the less noise is tolerable. In multi-story condominium projects where hard surface flooring such as marble, granite, ceramic tile, or hardwood is used, sound attenuation devices are a virtual necessity. The problem of floor-to-floor noise when hard surfaces are used is so acute that in many condominium developments the use of hard surface flooring is either prohibited or severely restricted.

In terms of room-to-room sound transmission, owners generally will not tolerate bathroom noises through party walls (in fact, such noise is virtually guaranteed to start a lawsuit for construction defects). Through-wall noise can be alleviated through a variety of means. Among them, careful sealing of the drywall at the bottom and top of the wall; careful sealing around electrical outlet openings, avoiding the condition where electrical outlets are lined up on both sides of a party wall; and the use of "staggered stud" construction on party walls, so that the drywall on each side of the party wall is not in contact with the studs of the adjoining room.

Floor-to-floor sound transmission can be attenuated by several means. The use of isomode-type hangers (a bracket assembly of steel and rubber used as a connection between the bottom of a ceiling joist and resilient channels), resilient channels, and special vinyl sheet membranes (made from extra heavy vinyl) can alleviate a substantial amount of sound transmission.

Noise from plumbing installation always results from a failure to isolate pipes. As with anything in construction, there are levels of sophistication, time trouble, and expense that can be applied to pipe noise. At a basic level, there is the choice between cast iron and ABS pipe. Although the UBC will sometimes require the use of cast iron pipe, ABS pipe is ubiquitous and, unfortunately, substantially noisier than cast iron. As noted, if either ABS or cast iron pipe, or copper supply lines, are not isolated from the building structure, noise will result. A more sophisticated (and expensive and time consuming) installation would use sound suppression pipe insulation, which must be installed in an exacting manner, including sealing all gaps and connections with appropriate caulking materials. This sort of installation provides a much higher degree of sound attenuation but is usually not feasible without an acoustical consultant and a willingness to plan the installation from the inception of the project (wrapping pipes with this material requires more space and access than in a more conventional installation).

Noise from hvac installations nearly always results from undersized ducts and air returns. In some instances, noise results from the failure to install noise isolation devices (isomodes) from roof and platform mounted equipment, which causes vibrations to occur within the structure of the building. Like plumbing issues, there are levels of sophistication, expense, and time in the soundproofing of hvac ductwork and equipment. It is entirely possible to wrap hvac ductwork, both inside and out, with sound attenuation devices that will reduce (or nearly eliminate) sound from the moving air. Again, this installation requires the services of an acoustical engineer and a commitment to an extremely detailed and methodical installation technique.

In areas where outdoor noise, such as traffic noise, is a problem, many builders automatically default to the use of double paned windows in an effort to alleviate noise. This is almost always a mistaken approach. Laminated glass windows are substantially more effective in reducing noise than double paned window assemblies.

Because noise attenuation is highly technical (though not often thought of in that manner by most builders) it is recommended that for any substantial (and certainly for any high-end) project that the construction professional retain the services of a qualified acoustical engineer.

At a basic level, here's how to avoid the defect:

- Pay careful attention to the assembly of party walls to minimize openings and avoid through-wall sound transmission.
- Obtain specifications for sound attenuation devices to reduce floor-to-floor sound transmission and follow such specifications to the letter.
- Confirm and verify that plumbing supply and drain lines have been isolated from the building structure.
- Do not undersize hvac supply and return lines or grills.
- Obtain a professional opinion with regard to exterior glazing for sound attenuation.

Defects in Wood Posts and Fences

Untreated wood posts that are in direct contact with earth or concrete will rot. Untreated wood posts, whether left exposed or covered with decorative elements, are commonly used to support entry roofs or porch structures. It is common to see such posts in direct contact with concrete or other hardscape (often the hardscape is installed after the post is installed and the level of the finished surface of the hardscape is 4 to 5 inches higher than the bottom of the post. Untreated wood in this condition is left exposed to repeated wet-dry cycles. Wet-dry cycles will cause rapid deterioration and dry rot of untreated lumber, requiring the replacement of the wood post structure and often the demolition of the surrounding hardscape.

Wrought iron and other metal fencing and railing elements are commonly installed as decorative or functional elements in exterior applications. The generic construction defect with such installations is a failure to prime the metal properly prior to painting. Failure of finished paint surfaces and deterioration from rust can occur within six months of installation, depending on climate conditions.

Here's how to avoid the defect:

- When installing wood posts, install a galvanized (or preferably stainless steel) post anchor to raise the bottom of the post at least 1 inch above finished hardscape surface or grade if installed in softscape.
- When installing wrought iron or other metal fencing materials, verify that the metal has been primed prior to the application of the finished paint surface.

Defects in Irrigation Systems

The subject of the installation of irrigation systems could well occupy an entire treatise, but for purposes of the most common defect involving such systems, the focus is narrow.

Irrigation systems that use sprinklers must be installed so that the spray from the sprinklers does not come into contact with the building surface, painted surfaces on wrought iron or wood fences, or hardscape that is subject to deterioration from water saturation.

Even a properly applied exterior finish system, whether brick, stucco, or wood siding, will deteriorate rapidly if exposed to constant water spray from sprinklers. Likewise, mineral deposits will deteriorate, or at least make unsightly, hardscape such as concrete or stone, and all painted surfaces, whether on metal or wood, will rapidly degrade from constant exposure to water from sprinklers.

To avoid the defect, verify that all sprinklers are adjusted so that they do not deposit water on the building structure, hardscape, or painted surfaces.

Defects in Dryer Ducts

Defects in the installation of dryer ducts not only cause customer complaints and litigation, they are potentially a life-safety concern. The build-up of lint in dryer ducts can cause the outbreak of fire.

The common defects in dryer duct installation are failure to install the duct inside the structure so that sufficient clearance exists to install the clothes dryer, and failure to install the duct in compliance with applicable codes and manufacturers recommendations and specifications, resulting in duct runs that are sloped improperly or too long to allow proper air movement out of the dryer to the outside.

All models of dryers require a connection between the dryer and the ductwork. This connection is typically made from flexible hose. Owners often expect (and architects often intend) that the dryer will sit flush, or nearly flush, with the wall. When the ductwork is not recessed into the wall cavity, the dryer must be moved outward into the room to accommodate the need for the connection between the dryer and the ductwork.

Of more functional importance is the slope and length of the duct run to the outside of the structure. Dryer ductwork that is angled upward will create a condition where lint builds up in the duct assembly. A substantial accumulation of lint can create a fire hazard. Duct runs that are too long prevent moisture from being ducted to the outside of the structure. At the least, the dryer will not function properly (taking too long to dry clothes).

Here's how to avoid the defect:

- Verify the installation of the dryer machinery and provide sufficient space for the connection of the dryer machinery and the duct to the exterior.
- Do not exceed the maximum length of duct run for the dryer duct.
- Verify that the duct has been sloped downward to the exterior.

MOLD

The net result of water intrusion into a building is not only the specific damage to drywall, paint, floors, and the like, but also the growth of mold and possibility of dry rot in wood framing members and other wood elements, such as trim and floors.

Figure 5-71 shows mold growth in a high rise (and high end) hotel room from water intrusion under a window that was improperly flashed.

It is not an understatement to say that mold has the construction, business, and real estate communities in a state of near panic. The seriousness of the problem is evidenced by, among other thing, the fact that in 2002, State Farm Insurance announced that it would no longer write new homeowner insurance policies in California, one of the most lucrative markets in the United States. Farmers Insurance no longer writes new homeowner policies in the state of Texas (perhaps because a jury issued

Figure 5-71

Mold growth from a window that was improperly flashed

a verdict against State Farm for $32 million in the famous Ballard bad faith mold case—which was subsequently reduced on appeal to a more modest number). Since the start of 2002, dozens of commercial lines carriers have refused to write new policies on commercial buildings and apartments, leaving a critical shortage of carriers and coverage.

Equally troubling is the mystery surrounding mold, the lack of information and the misinformation about mold, the effects of mold on the health of building occupants, and the practical and legal ramifications for building owners. In short, building owners and contractors are experiencing enormous difficulty in getting their arms around this problem.

How and Why Does Mold Grow?

Mold is a type of multicellular fungus that grows as a mat of microscopic filaments called *hyphae* and is dependent on an external food source, because, like all fungi, it has no chlorophyll to manufacture food from internal photochemical processes.

Mold forms when excess moisture is present inside a building, where a food source is available, and where the temperature is at least 50°F. Generally, if the moisture content of the material upon which the mold might be growing is less than 20 percent, the mold will not grow.

The prime surface for mold growth is cellulose, or wood fiber. However, mold can grow on a variety of surfaces, and some types can even grow on dust. Drywall is an excellent food source for mold because a sheet of drywall is a sandwich of gypsum covered on the front and back with paper. Paper, of course, is made from cellulose (wood fiber). For mold, the paper coating on a sheet of drywall is virtually a predigested dinner.

Moisture intrudes in buildings in one of two ways: it leaks through doors, windows, and roofs; and it floods from leaking or burst pipes. When the temperature and humidity are high enough, mold can begin to grow in as little as 24 to 48 hours. When owners are unaware that water is intruding into the building (a slow leak from a hidden pipe, for example), the mold has ample time to grow and may spread dramatically before it is discovered. When the owner sees the water in the building (from a burst pipe, causing a flood), owners may clean up and dry out the building, but they may be unaware that enough moisture has been present that mold has started to grow before the building was cleaned up and dried out. The same holds true for moisture from a leaking roof or leaks through windows and doors. If the building is warm enough, and if there is food for the mold, mold can and usually will grow within 48 hours. Merely drying out the building will not remove this new mold growth.

What Is Toxic Mold?

Not all mold is considered toxic, even by those who believe that there are some forms of toxic mold, and there are many people who reject the notion that mold is toxic in any manner. There is little consensus in the medical community on this issue. Perhaps the most reliable information as of the date of this publication is the 2002 position statement from the American College of Occupational and Environmental Medicine (www.acoem.org). In this position paper, the ACOEM notes that mold affects human beings in three possible ways: (1) through allergic reactions; (2) by direct infection, usually in open wounds; and (3) by toxicity, which in this context means the release of a colorless, odorless gas called a mycotoxin. Of the three possible manners in which mold may affect humans, allergic response is by far the most common.

With regard to allergic reaction, ACOEM first notes that only that portion of the population that is *atopic* will have the chance to suffer an allergic reaction to mold. Atopic individuals have allergic responses to a wide range of allergens. Persons who are not atopic are theoretically not capable of experiencing an allergic reaction to mold. Of the population of atopic individuals, which is about 40 percent of the general population, only about 25 percent of atopic individuals (meaning 10 percent of the whole population of both atopic and non-atopic individuals) will suffer allergic reactions to molds.

Generally, allergic reactions to molds will manifest themselves as allergic asthma or allergic rhinitis (sneezing, runny nose, or other cold-like or hay-fever-like symptoms). Rarely, and usually only upon exposure to very high levels of certain molds, more serious respiratory disease develops. Most allergic symptoms disappear when the mold is removed from the area around the person or the person is removed from the area in which the mold is present.

The ACOEM notes a significant reality about mold with respect to allergic reactions. Because mold is ubiquitous in the environment, meaning that it exists everywhere, and in particular both inside and outside of structures, an atopic individual who is susceptible to allergic reactions may develop such a reaction to mold in the outdoor environment and not merely from indoor mold. In other words, not every allergic reaction to mold is caused by mold in the indoor environment, and many such reactions are caused by mold in the outdoor environment.

As a general proposition, the remedy for allergic reaction to mold is to eliminate the source of the indoor mold, clean mold spores from the indoor environment by the use of air filtration, minimize the intrusion of mold spores from outdoor mold into the structure using air filtration, and lower humidity levels.

Infection is the second type of risk from mold. Serious fungal infections from deep wounds generally only occur in severely immunocompromised individuals, such as cancer or AIDS patients. Superficial fungal infections are common, however, such as athlete's foot, or fungal infections of the toe and fingernails. The group of fungi responsible for these type of infections, although they live indoors, are generally not thought of as being caused by water intrusion into structures.

The third, and most controversial, type of possible harm to human beings is toxicity. In this context, *toxicity* means injury from mycotoxins, which are metabolites that are produced by molds as they grow and as they die. Metabolites are colorless, odorless gases. They are widely reported in the media as having the same or similar chemical composition as nerve gas, and some people contend that they have neurological effects on human beings, such as headache, dizziness, and loss of short-term memory.

Conceptually, with regard to the issue of toxicity, the first important thing to understand is that not all species of mold produce mycotoxins. As a general matter, even for those proponents of the concept of toxic mold, molds that do not produce mycotoxins are not considered toxic for the purpose of toxicity. Molds that do not produce mycotoxins may, however, still produce allergic reactions in atopic individuals.

Mycotoxins themselves are large molecules that are not volatile, meaning that they do not readily evaporate into the atmosphere. Not only do they not readily evaporate, they also do not off-gas, like emissions from solvents or organic chemicals. Mycotoxins make their way into the atmosphere by attaching themselves to other particles that in turn make their way into the air. Thus, to inhale a mycotoxin, an individual would have to inhale a piece of mold, whether in the form of the substrate to which the mold is attached (dust, wood fiber), a fragment of the mold colony (a strand or piece of hyphae), or a mold spore, to which the mycotoxin would be attached. The mycotoxins themselves are colorless and odorless.

Many people determine that mold is present through its easily recognizable musty odor. This musty odor is not produced by mycotoxins. The common musty smell of mold is instead caused by a different set of compounds, called *microbial volatile organic compounds* (MVOCS), that have a low molecular weight, making them volatile and easily dispersed into the atmosphere.

The question is whether the mycotoxins cause adverse health effects in humans. Of the molds that produce mycotoxins—i.e., the so-called "toxic molds," the most publicized is certainly *Stachybotrys chartarum*. It is often argued that other toxic molds include *Aspergillus versicolor* and *Pennicillium*. Articles in newspapers and magazines invariably refer to *Stachybotrys* as the main culprit responsible for serious illness. One reason is perhaps the now infamous CDC infant study. Following an investigation in 1993 and 1994, the Centers for Disease Control initially surmised that the deaths of a few very young infants may have been caused by *Stachybotrys*, but the data from this study was later reviewed and the CDC concluded that there was no causal relationship between the exposure to mold and the injuries suffered.

After an extensive peer-reviewed analysis of the evidence and data, the ACOEM concluded the following in its 2002 position paper with regard to the so-called toxic molds:

> "Some molds that propagate indoors may, under some conditions, produce mycotoxins that can adversely affect living cells and organisms by a variety of mechanisms. Adverse effects of molds and mycotoxins have been recognized for centuries following ingestion of contaminated foods. Occupational diseases are also recognized in association with inhalation exposure to fungi, bacteria, and other organic matter, usually in industrial or agricultural settings. Molds growing indoors are believed by some to cause building-related symptoms. Despite a voluminous literature on the subject, the causal association remains weak and unproven, particularly with respect to causation by mycotoxins. One mold in particular, *Stachybotryos chartarum*, is blamed for a diverse array of maladies when it is found indoors. Despite its well-known ability to produce mycotoxins under appropriate growth conditions, years of intensive study have failed to establish exposure to *S. chartarum* in home, school, or office environment, dose-response data in animals, and dose-rate considerations suggest that delivery by the inhalation route of a toxic dose of

mycotoxins in the indoor environment is highly unlikely at best, even for the hypothetically most vulnerable subpopulations.

Mold spores are present in all indoor environments and cannot be eliminated from them. Normal building materials and furnishings provide ample nutrition for many species of molds, but they can grow and amplify indoors only when there is an adequate supply of moisture. Where mold grows indoors there is an inappropriate source of water that must be corrected before remediation of the mold colony can succeed. Mold growth in the home, school or office environment should not be tolerated because mold physically destroys the building materials on which it grows, mold growth is unsightly and may produce offensive odors, and mold is likely to sensitize and produce allergic responses in allergic individuals. Except for persons with severely impaired immune systems, indoor mold is not a source of fungal infections. Current scientific evidence does not support the proposition that human health has been adversely affected by inhaled mycotoxins in home, school, or office environments." (American College of Occupational and Environmental Medicine, Evidence Based Statements, 2002. "Adverse Human Health Effects Associated with Molds in the Indoor Environment.")

How Is Possible Mold Contamination Confirmed?

Whether or not you are convinced that mold is actually dangerous, just about everyone in the construction industry would agree that it is important to find out if mold is present, and if so, to remove it.

To determine whether mold exists in a building, tests should be conducted by a certified industrial hygienist. Two kinds of tests are made. First, where mold is visible, the hygienist will take a physical sample, with tape or a cotton swab, and the suspect material will be placed into a growth medium in a petri dish. Second, the hygienist will take air samples, both inside the building and outside the building. The air sample is captured onto a dish containing a growth medium (similar to a petri dish). Both of the samples on the dishes are then sent to a lab. After about a week, enough mold will have grown inside each dish so that it can be analyzed. The analysis of the surface sample is intended to identify the type of mold. The analysis of the air sample is intended to identify the type of mold and the number of spores per a given area (this is called a *spore count*). A report will then be prepared by the laboratory and by the hygienist. With regard to the spore counts, the report will compare the spore counts inside and outside the building. Because mold is everywhere in the environment, generally speaking a building is "contaminated" by mold only when the spore counts inside the building are substantially greater than the spore counts outside the building.

What Happens if Mold Is Present?

Many people contend that if mold is found in the structure, the occupants should immediately move out until the mold is removed. There is a wide difference of opinion on this topic. Some medical professionals believe that if toxic mold is found in any significant concentration inside a building, the tenants should move out, particularly if their immune systems are compromised or if small children are living in the structure. However, some medical professionals believe that it is not necessary to move out of the building, particularly if the occupants are not showing any symptoms or problems that might be caused by the mold.

There is no question, however, that mold must be *remediated*, which is the technical term in the industry for removal or eradication of mold—or in plain English, *cleaned up*. Even in situations where the occupants may not show physical symptoms, it is probably better practice to remediate the mold to avoid the chance of allergic reaction. In addition, because the presence of mold must be disclosed to a buyer of real property, from a purely economic point of view, it makes sense to remove mold as soon as possible. From the perspective of an owner of commercial property, such as an apartment or office building, taking into consideration potential legal liability predicated on basic notions of reasonable conduct, remediation should begin at once.

How Is Mold Remediated?

Mold is remediated by a mold remediation contractor. At this point, a word of caution is in order. Many contractors, seeing a potential bonanza in mold remediation, hold themselves out as qualified mold remediation contractors, when in fact they have only limited experience and understanding.

Mold is remediated in the following manner. First, the source of the water intrusion must be located and eliminated, because mold cannot grow without moisture. Obviously, it makes no sense to eliminate the mold inside the building without finding the source of the water intrusion, because if moisture continues to intrude into the building, new mold will grow as soon as the old mold is removed.

Second, the remediation effort then focuses on the type of material that has been contaminated. The method of remediation is different for each type of material. Basically, three types of material are in question. The first type is porous material, such as upholstered furniture, carpets, rugs, linens, clothing, and books. Depending on the degree of contamination, some of these items can be cleaned using a special vacuum called a "hepa vac" (a vacuum that has a special filter that traps the mold spores inside the vacuum system so that they are not spread into the atmosphere during the vacuuming process). Some contaminated porous materials will have to be discarded.

On the subject of clothes, it is generally accepted that clothes that can be laundered should be laundered in hot water with bleach, if possible. Not all clothes can be laundered. On the issue of dry cleaning, there is no generally accepted consensus in the mold remediation industry that dry cleaning will destroy mold spores. Some remediation contractors recommend that clothing and other porous items be exposed to high concentrations of ozone. An ozone generator can be placed inside a house (with appropriate sealing of doors and windows) to expose furniture, carpets, rugs, wall coverings, clothes, and other porous items to the ozone. Some commercial dry cleaning establishments offer ozone treatments at their plants. Again, however, there is no generally accepted consensus that exposure to ozone has any substantial impact upon mold spores.

The second type of item is hard-surfaced items, such as mirrors, glass, and countertops. To clean these items, they are "wet-wiped" with a fungicidal solution. In this case a *fungicidal solution* usually means a mixture of water, detergent, and 10 percent common household bleach. The detergent component is important because mold spores have a waxy covering that makes them resistant to chemicals, even bleach, and the detergent helps to break through this waxy surface coating. Hard-surfaced items are the simplest to clean.

The third type of item is drywall. In concept, remediation of mold on drywall is relatively simple: one removes the drywall and discards it. In practice, however, remediation of drywall is time consuming, difficult, and expensive for a number of reasons. First, if mold is growing on drywall, the mold spores will be disturbed during the process of removing and discarding the drywall. When the mold spores are disturbed, they will float into the atmosphere and can either be inhaled by the residents or workers or be deposited in other parts of the structure. If there is sufficient food and moisture available where the mold spores land, it is possible that a new spore-producing colony will begin to grow at the new location. For this reason, it is important to ensure that during the process of removing drywall the mold spores are not allowed to migrate to other parts of the house.

The process of isolating the mold spores during removal of contaminated material is called *containment*. The process of removing mold spores using containment techniques is called removal *under containment conditions*. Containment conditions usually involve the following technique: (1) a fairly large machine, about the size of a chest of drawers, called a *negative air machine* is brought into the room. This machine is basically a large vacuum that sucks air from the room into a duct, or flexible tube. The duct is run from the machine to the outside of the house (either a hole is cut into an outside wall, or the duct is placed through a window and the area around the duct sealed with plastic and tape); (2) the area to be contained is sealed off from the rest of the structure, using plastic and tape, so that the air inside the contained area is pulled out from that area to the outside of the house; (3) an air-tight entry is made of plastic to allow workers to enter and exit the contained area.

Once the area is contained, the remediation contractor needs to determine what portion of the drywall is contaminated. When the source of moisture is a flood that releases water onto a floor, the standard protocol is to remove the baseboard and remove the drywall on the walls from the floor up 4 feet on the wall. When water has intruded into the ceiling, the drywall on the ceiling is removed first, and holes are cut into the drywall on the walls to determine whether any contamination exists on the walls. If so, the drywall on the walls is removed as necessary.

The drywall itself should be bagged in plastic bags that are then carefully sealed, removed from the contained area, and discarded so that mold spores from the contaminated drywall do not spread to other parts of the structure.

Once the drywall is removed, a dehumidifier is introduced into the contained area. This machine draws moisture out of the area, including the wood framing and any drywall that may be left on the walls. Once the area has been thoroughly dried, the exposed framing is scrubbed with stiff brushes, using a solution of water, detergent, and 10 percent bleach. Once the wood framing is dry, the wood framing is vacuumed using a hepa vac.

Once the exposed wood framing has been cleaned, it is sprayed with a biocidal coating. This application to the exposed framing is intended to accomplish two things: (1) the material "encapsulates" the framing, the theory being that it will "seal" the wood and prevent any spores from entering the room; and (2) the biocidal chemicals in the product will kill any new spores that might grow on the framing.

There are a few important aspects to this process. First, the wood must be clean and dry before the coating is applied. Second, the coating must be applied to a certain minimum thickness (measured in a certain number of "mills" or thousandths of an inch). Care should be taken to insure that the minimum recommended thickness is actually applied by the remediation contractor.

One final note on the biocidal coating before leaving the subject. There are some construction professionals who believe that particular attention should be paid to mold even during the rough framing phase of new construction. These professionals recommend that common "lumberyard mold" (a black powdery mold that grows on wet wood at the lumber mill, during shipping and during storage at the lumber yard) should be removed by scrubbing with a brush and 10 percent bleach and detergent solution. Some of these professionals also believe that the framing in all areas that will be exposed to high levels of moisture—i.e., bathrooms and kitchens—should be coated with biocide immediately after framing.

My own view of this approach is that while some people might regard it as overkill, I endorse the careful approach. The removal of lumberyard mold during and immediately after framing, even in a large home, requires only a few hours work by a common laborer. The application of biocide requires somewhat more work, but the material will not add substantially to the cost of the home, and the amount of time that it takes to spray the material on exposed framing is minimal. In this regard, the large mold/personal injury case that was settled in the Los Angeles County Superior Court in November 2005 involved a claim of lumberyard mold. Although there are many reasons, some procedural that are not related to the merits of the plaintiff's underlying claim in that lawsuit, the fact is that the lumberyard paid approximately $10 million to settle the plaintiff's claim (others involved paid collectively about $13 million). To my knowledge, this is the first reported instance in the United States where a payment was made on account of claims arising from lumberyard mold. Whether this has an impact upon the common handling of lumber and the installation of framing lumber on jobsites remains to be seen.

Once the wood framing has been cleaned and treated, new drywall is installed, taped, and finished. One of the reasons why the remediation of contaminated drywall is so expensive is the containment process. The hepa vacs and the negative air machines are expensive. The remediation contractors usually rent them on a daily or weekly basis. When a mold remediation job is done carefully, the workers will wear protection gear, consisting of overalls, gloves, and respirators. These outer garments should be discarded in a fashion to restrict the spread of mold spores to other areas of the structure. When one combines the cost of the machines with the cost of the "consumables"—i.e., the overalls and other protective gear that has to be discarded—the costs become quite substantial.

After the remediation is completed, the hygienist re-tests the building by taking air samples to determine whether the mold is gone. If the lab reports show that the mold has been removed, the hygienist then issues a *clearance*, which is the official confirmation that the remediation effort has been successful.

Legal Issues Involving Mold

The legal issues include the following:

- Potential liability on the part of the owner of a commercial structure such as an office building or apartment for exposing occupants of a building to toxic mold, including claims for bodily injury, lost wages, and emotional distress.
- The availability of insurance for first-party claims.
- The availability of insurance for third-party claims.

There is little doubt that a building owner can be sued for exposing occupants of a building to an unsafe environment, although the proof of such a claim is difficult, and the injuries are difficult to quantify. The plaintiff in a lawsuit against a building owner would claim, and if successful, be entitled to damages for medical expenses, pain and suffering, lost income, and possibly emotional distress if the plaintiff could prove that the owner acted unreasonably in failing to maintain the building or remove the mold, and that the plaintiff's injuries were caused by the mold.

The issue of causation (i.e., is the mold the cause of the plaintiff's injuries?) is of major concern in both the medical and legal communities. Recent cases at the federal level have raised the standard for the admissibility of expert testimony, and those cases require that the expert's opinion be generally accepted in the expert community (this is called *peer acceptance*). When mold is at issue, the plaintiff may have a difficult time having the courts accept expert testimony to establish the casual connection between the mold and the injuries. In recent cases that have gone to trial, the plaintiffs have, in large part, relied on empirical evidence to prove their case. As a general matter, a plaintiff's case will stand or fall on expert testimony. If the court decides to disallow the plaintiff's expert from the testifying, because the court is not convinced that the medical community generally accepts that there is a connection between exposure to mold and physical injuries, the plaintiff's case is over.

The insurance picture on mold problems is somewhat fact dependent. In the area of first-party claims, that is, a claim by a building owner against his or her own insurance carrier, the analysis begins with the insurance policy. Carriers typically resist mold claims in several ways. First, they argue that the source of the water intrusion causing the mold is not a covered claim. This argument is always made when the water comes from a leaking roof, or results from poor maintenance. Second, the policy may have a mold exclusion. Because the carriers are so well aware of the enormous cost of mold remediation, they have exhibited strong resistance to paying mold related claims.

In the area of third-party claims the picture is somewhat different. Third-party claims in this area usually arise in the context of a construction defect claim. In that case, the mold, and the cost of remediation is usually analyzed in the context of "resultant damage," a term of art referring to damage that is customarily covered under Commercial General Liability Policies.

To a certainty however, it will be virtually impossible to obtain a new policy for a commercial building or apartment with mold coverage, because the insurance industry has, in the last few years, established a national database for mold and water intrusion claims. To put it bluntly, a building owner has a Hobson's choice: he or she can press the insurance carrier for payment on a claim and possibly succeed, but almost certainly jeopardize the possibility of renewing the policy or obtaining another policy, or simply decide to pay the expensive damage without a contribution from the insurance company.

Protecting Against Mold

First, find and fix any areas of the property where moisture is intruding into the building. Look carefully at your maintenance program, including roof inspections. Second, if you find water coming into the building, fix the problem. Third, if you have found water in the building, have the building tested for mold. Fourth, if you find mold in the building, remediate it without delay. Fifth, if you are the owner of commercial property, such as an office building or apartment building, consider moving occupants out of the building, particularly if the occupants complain of physical symptoms. Sixth, if you find mold, notify the occupants so that they cannot claim that you concealed the existence of mold from them.

6 Using Plans, Specifications, and Contract Terms to Avoid Defects

FOR OWNERS AND GENERAL CONTRACTORS

In a very real sense, avoiding construction defects is a function of understanding what defects usually occur, understanding why they occur, and paying attention to construction details. Usually, it's the paying attention part that causes problems.

There is no single best way to insure that workers are paying attention to the details. Traditionally, contractors paid attention to details through supervision. Supervision is, and always has been, an essential component of an overall strategy to achieve better results in construction. But effective supervision depends in substantial part on the quality and knowledge of the supervisor and on the supervisor's ability to remember the hundreds of details that go into any construction project.

I believe that plans, specifications, and specific language in contracts can also assist contractors, subcontractors, and owners in avoiding construction defects.

If you are a contractor, from the point of view of avoiding litigation, plans and specifications are the most overlooked documents in any construction job. But plans and specifications can be used to avoid lawsuits, to make it easier and cheaper to defend them if they occur, and to make sure that the people responsible (and not you) have to pay the damages.

If you are an owner, the effective use of plans can help you create a better building and help insure that contractors and subcontractors follow the plans, follow the specifications, follow the manufacturer's recommendations, and obtain answers to questions that might arise during construction instead of simply ignoring them and using their own (many times incorrect) solutions to plans that are confusing or inadequate.

A bit of overview is necessary to understand how plans and specifications can be used to avoid litigation and create better buildings. First, thoughtful use of the plans and specifications are important more for avoiding construction defect claims than for avoiding claims involving delay and cost overruns. Second, in virtually every construction defect claim, the owner will contend that the plans or specifications were not followed. Third, in many cases, the plans and specifications will either be ambiguous, or there will be a debate over whether a particular contractor or subcontractor was made aware of the plans and specifications and had an obligation to follow them. Fourth, the subcontractors will almost always argue that the plans were inadequate, conflicting, or not otherwise suitable. Fifth, the subcontractors will argue that while not perfect; their work was "good enough."

With this in mind, plans can be particularly useful in several ways as a means to avoid litigation and to ensure a better built project. To avoid the argument down the road from a subcontractor that he or she wasn't aware of what the plans required, you can request that the architect insert a signature block on the plans themselves. That signature block will provide space for you, as the contractor, and all the subcontractors, to sign and acknowledge the following:

- That you and they have read the entire set of plans
- That you and they understand the entire set of plans
- That the plans are adequate to complete the work
- That the plans do not contain any discrepancy or ambiguity that would interfere with the successful completion of the work
- That you and they will strictly follow and comply with the plans and specifications
- That any variance from the plans and specifications must be approved in writing
- That any inadequacy in the plans must be brought to the attention of owner and contractor in writing

Second, you can request that the architect insert into the plans a great number of particulars to establish standards and practices for yourself and the various subcontractors. These could include the following

- Any particular means or method of building the project
- Any particular standard of workmanship
- Any particular material to be used

What have you accomplished by placing this language in the plans themselves? First, you have completely eliminated the argument that a particular subcontractor did not see the plans, because you have a set of plans signed by htat subcontractor acknowledging that they were read, reviewd and understood. Second, you have made it harder for the subcontractor to argue that the plans were confusing, abiguous, inadequate, or insufficient, because the subcontractor has acknowledged in writing the plans suffered from none of those problems. Third, you have made it more difficult for the subcontractor to raise the defense that although the subcontracor did not comply with the plans the work is nevertheless adequate, because the subcontractor has acknowledged in writing that he or she was required to comply strictly with the plans. In short, you have given yourself much greater protection with respect to all of the common defenses raised by subcontractors in construction defect lawsuits.

By placing the language in the plans, and not in the contracts with the various subcontractors, you also avoid the issue of any conflict between the language in those subcontracts and the plans themselves. Thus, for example, a particular subcontractor might be contractually required to install a building component in a particular way. If that requirement is only in the contract, it gives the subcontractor a good fall-back argument that he or she could not comply with the contract because the contract varied from the plans. Putting the particular language in the plans themselves gets rid of that problem.

Owners and architects might be unfamiliar with the idea of using the plans in this fashion, and architects in particular might be resistant to the idea. In addition, you (the contractor) might be hired on the job after the plans are fully complete and you might not have the opportunity to insert your desired language. In those circumstances, another approach will suffice. Include your language in the contract with the subcontractor, and attach the plans and specifications as an exhibit to the subcontract. Have the subcontractor initial each page of the plans and keep a copy of the subcontract with the attached plans (bearing the subcontractor's initials) in your files.

All building plans contain general notes. The general notes are sometimes just that: general and usually unhelpful in the real world to avoid construction defects. But anti-construction defect language can be inserted into the general notes of the plans and inserted in such a way as to insure that the subcontractors read and understand the general notes and the specific provisions that apply to their work.

What should be inserted into the general notes? Following is "anti-construction defect" language for general contractors and for some of the main subcontractors. This language could be inserted into the general notes on the plans or it could be used as an addendum to the contracts between the owner and contractor and contractor and subcontractors.

OWNER/CONTRACTOR AGREEMENT SPECIAL ANTI-DEFECT PROVISIONS

The following language is somewhat general. Some of the provisions are not applicable to all residential or commercial construction.

Special Requirements of Contractor with Respect to Specific Building Conditions and Components

Contractor shall use every effort to insure that the subcontractors comply with the following special requirements of the project:

With respect to the below-grade walls, including retaining walls,

- The Subcontractor shall strictly comply with all recommendations of the waterproofing consultant with respect to application of the waterproofing materials on the outside of the masonry walls.
- The Contractor shall confirm that protection board or equivalent has been placed against the waterproofing materials prior to backfill or installation of perforated drains.

- The Contractor shall verify that the proper type and amount of crushed rock has been placed against the below-grade waterproofed walls.
- The Contractor shall verify that the perforated drain pipe has been installed with the perforations placed on the bottom.
- The Contractor shall verify that filter cloth has been placed around the perforated drain pipe and over the crushed rock before backfill.
- The Contractor shall verify that the perforated pipe has been sloped to drain and properly connected to a drain assembly that daylights or enters a sump.
- The Contractor shall confirm that the backfill against the below grade walls is compacted to at least 90 percent relative compaction and that the backfill material is free from rock, debris, and organic material.
- With regard to the drain at the bottom of the driveway leading to the underground garage, the Contractor shall confirm that no construction debris has been left in the drain assembly after completion and the Contractor shall water test the drain assembly.

With regard to the application of the exterior plaster system (three coat stucco system), Contractor shall confirm the following:

- That all plywood shear panels or exterior sheathing have been chalk-lined to show the location of studs for installation of lath.
- That all lathing material is stapled or nailed to studs and not sheathing.
- That the requisite length of fastener has been used to attach the lath to the structure.
- That two layers of Type D builder paper or equivalent has been used to wrap the building (no K-Lath or equivalent shall be used to lath the building).
- That all builder paper is properly shingled and not reverse lapped.
- That all builder paper butts against the exterior window frames as required by code.
- That the builder paper is free from rips and tears.
- That no builder paper is applied on any horizontal surface.
- That the scratch coat is properly scratched to provide a mechanical bond to the brown coat, that it is sufficiently keyed into the lath, and that it is watered as necessary after installation to hydrate the plaster properly.
- That all cracks in the scratch coat are repaired before installation of the brown coat.
- That the brown coat is watered as necessary after installation to hydrate the plaster properly.
- That any cracks in the brown coat are repaired before installation of the color coat.
- That any below grade masonry or concrete walls that will be stuccoed shall have a positive side waterproofing system, with a perforated drain system.
- That the brown coat is applied not less than 14 days after the scratch cost is applied, and the finish coat is applied not less than 14 days after the brown coat.
- That all weep screeds have been installed and that such are at the proper height above grade as called for in the applicable building code and free from excess stucco that might impede the drainage of water.

With regard to the rough framing,

- That all hold-downs and anchor bolts are securely fastened upon initial installation and that all such devices are re-tightened prior to the application of exterior sheathing or interior drywall.
- That all walls and floors are checked for square plumb and level.
- That all walls and ceiling joists are straight-edged prior to installation of drywall.
- That all sawdust and wood chips are removed from stud bays prior to installation of drywall.
- That all fire blocking has been installed per plans.
- That all metal structural devices, such as hold downs, clips, or other anchors that directly contact pressure treated wood, are of a composition to resist deterioration from the chemicals in such wood.
- That all full height perimeter blocking, if any, is in direct contact with roof sheathing.
- That all visible "lumberyard mold" be removed by scrubbing the same with an application of a solution of water, detergent, and a minimum 10 percent chlorine bleach before installation of interior drywall and exterior sheathing.

With regard to shear panels,

- Contractor shall personally inspect the type of nails used to nail the shear panels to determine whether they conform to the structural engineer's requirements.
- Contractor shall personally inspect the nailing of the shear panels to confirm that:
  A. Nails are not over-driven.
  B. Nails are not driven closer than 1/4 inch from the edge of any plywood sheet.
  C. Nailing conforms to the specified nailing patterns.
  D. Shear panels have a 1/8 inch gap between each panel.
- Contractor shall reject and require the removal and installation of a new shear panel for any shear panel that does not strictly conform to the above requirements.
- All shear walls shall be numbered on the plans and a corresponding number spray painted onto the shear wall in the field. Contractor shall verify that all shear panels called out on the plans have been installed.

With regard to party walls, Contractor shall verify that the drywall installation conforms to the plans and specifications.

With regard to the installation of rough plumbing,

- Contractor shall verify that the cut ends of all copper pipe have been reamed. Contractor shall conduct an inspection, without notice to the plumbing subcontractor, to determine whether the copper pipe has been reamed. If Contractor determines that the pipes have not been reamed, Contractor shall require the plumbing pipes to be removed, reamed, and reinstalled, at the subcontractor's expense.
- Contractor shall confirm that all pipe has been properly isolated from the framing of the building.

- Contractor shall confirm that all dielectric fittings required by the uniform plumbing code have been installed.
- Contractor shall verify that all penetrations of all studs, beams, joists, and plates have not compromised the structural integrity of such assemblies.
- Contractor shall verify that all drain, waste, and vent (DWV) ABS lines are properly isolated from the framing to reduce noise.
- Contractor shall verify that all angle stops have been installed at the correct heights and locations.
- Contractor shall verify that all protective wall plates have been installed in areas where plumbing pipe runs in stud walls.

With regard to the installation of the rough electrical system,

- Contractor shall confirm that no penetrations of plates, studs, beams, or joists compromise the structural integrity of such assemblies.
- Contractor shall take steps to protect electrical wiring and electrical panels from drywall joint compound and paint over spray.
- Contractor shall confirm that all electrical circuit breaker panels are clearly and completely labeled.

With regard to penetrations of the building envelope by pipes and ducts,

- Contractor shall verify that all penetrations have been adequately flashed before application of the building paper, using best methods.
- Contractor shall verify no penetrations are cut after application of the building paper.

With regard to waterproofing around windows and doors,

- Contractor shall confirm that the subcontractor responsible for the installation of flashing has strictly and completely complied with the recommendations of the waterproofing consultant and the manufacturer and that all window and door openings have been properly and completed flashed prior to installation of door and window assemblies.
- Contractor shall confirm that bituminous peel-and-stick membrane flashing (such as Bithuthene or equivalent) shall not be exposed to UV light for longer than recommended by the manufacturer.
- Contractor shall confirm that when the waterproofing consultant and manufacturer recommend use of primer under peel-and-stick membranes that such has been done.
- To the extent that any window or door has a factory installed flashing, such as Moistop, Contractor shall confirm that such flashing is not removed in the field.

With regard to the installation of continuous sill pans, Contractor shall confirm that sill pans are infact continuous, that the vertical legs of such sill pans turn upwards an adequate amount, and that no nail, screw or other fastening device penetrates such sill pan after installation.

With regard to the framing of balcony decks and the waterproofing of the same,

- Contractor shall personally verify in the field that all balcony decks slope away from the structure of the building and that they are adequately sloped to drain.
- Contractor shall verify that all *L* metal flashing is continuous and properly installed.
- Contractor shall prevent injury to any waterproof membrane after installation by protecting the same from damage from nails, screws, scaffolding, extreme exposure to the elements, and the like.
- Contractor shall confirm that all sheet metal scuppers are set at the proper level.
- Contractor shall personally confirm that the secondary drain holes in all double hub deck drain assemblies are free from blockage and debris.
- Contractor shall verify that all balcony parapet walls have been properly flashed with either Ice and Water Shield, Vycor Ultra or equivalent, or a sheet metal coping cap with proper laps to prevent water intrusion into the parapet wall framing.
- Contractor shall prevent damage to the waterproof membrane during installation of the final finished balcony surface.

With regard to the installation of roofing material,

- Contractor shall verify that the requisite number and spacing of fasteners has been used to attach the roof underlayment.
- Contractor shall water test the roof after installation to confirm that no areas of the roof pond water.

With regard to drywall installation,

- Contractor shall verify that all plumbing protective plates have been installed.
- Contractor shall verify the spacing of drywall screws.
- Contractor shall verify that drywall nails are not used to attach the drywall to the framing permanently.
- Contractor shall verify that all corner beads are adequately taped to prevent cracking and that all corner beads are straight, plumb, and square.
- Contractor shall verify that all areas where two layers of drywall are required have two layers of drywall installed.
- Contractor shall verify that all one-hour or other rated assemblies comply with the plans and Code.
- To the extent that the acoustical engineer requires resilient channel or other sound-attenuating devices, Contractor shall confirm that such have been installed.

With regard to the elevator installation, Contractor shall insure that water does not accumulate in the elevator pit.

With regard to the installation of the hvac system,

- That the hvac contractor balances the system to confirm that the system adequately heats and cools individual units and public areas.
- That all ductwork is free from leaks.

- That all roof mounted equipment is mounted on isomodes and that all seismic tie-downs are properly installed.
- That all roof mounted ductwork is properly isolated from the roof membrane.
- That all roof mounted ductwork is configured so that it sheds water and does not pond water.
- That interior duct runs are not crushed, bent or do not otherwise impede the flow of air.
- That all hvac penetrations of the roof and walls are properly waterproofed. No hvac refrigerant line shall be placed directly through an opening in aroofunless same is installed using a "gooseneck" roof jack. No hvac refrigerant line shall be placed directly through a roof or wall opening where such opening is only sealed with foam around the refrigerant line.

With regard to the installation of shower pans and shower enclosures,

- Contractor shall prevent damage to or penetration of hot-mopped shower pans by construction debris or construction activity.
- Contractor shall verify that a monolithic waterproofing system has been installed on all vertical walls of shower enclosures and that such system is not penetrated or damaged by subsequent construction activity, including the installation of lath for a mortar bed for tile on shower enclosure walls.

With regard to concrete flatwork, Contractor shall verify that flatwork conforms to ADA (Americans with Disabilities Act) accessibility requirements and flatwork does not pond water (Contractor shall confirm this by water testing).

With regard to wrought iron installations, Contractor shall verify that all wrought iron is free from rust and that it is properly and completely primed and painted and that no landscape sprinkler is adjusted so that water is sprayed on wrought iron installations.

With regard to the installation of any concrete that requires welded wire mesh as a reinforcing material, Contractor shall verify that the welded wire mesh is supported by chairs, dobies or other devices so that the welded wire mesh is embedded in the middle of the pour, not at the bottom of same.

With regard to the installation of any water vapor barrier under concrete, Contractor shall verify that such vapor barrier is present, continuous, and not ripped or otherwise damaged before installation of concrete.

Contractor is advised to consider the use of Densgold, Densglass, or equivalent instead of greenboard in bathrooms and other wet environments, and the use of Fosters biocidal encapsulent paint on the framing of baths and other wet areas.

SPECIAL ANTI-DEFECT CONTRACT PROVISIONS FOR SUBCONTRACTORS

The following are suggested anti-defect contract provisions for the most important subcontractors.

Anti-Defect Provisions for the Window and Door Installer

Addendum to contract for door and window installer

This Addendum is made a part of and is incorporated into the contract between _____ Construction (General Contractor) and _____, (Subcontractor).

In the performance of Subcontractor's work on the Project, Subcontractor shall do the following:

1. Subcontractor shall strictly comply with the requirements of the Owner's waterproofing consultant with respect to the installation of all door and window assemblies.

2. Unless otherwise specified by the plans, specifications, manufacturer's recommendations and requirements, and recommendations of Owner's waterproofing consultants:
   A. All door and window assemblies shall be flashed with an appropriate above-grade peel-and-stick flexible rubberized asphaltic membrane (W.R. Grace Ice & Water Shield or equivalent), which membrane shall be primed, properly lapped, and not exposed to UV light for more than the time specified by the manufacturer. All such flashings wrap into the door and window openings. No sisal kraft type flashings shall be used.
   B. All door and window assemblies shall have a continuous sill pan of either extruded PVC (i.e., Jamsill or equivalent) or copper or leaded copper. No galvanized window or door sills shall be used. The sill pan under windows shall be shimmed with plastic shims so that it provides positive drainage to the outside of the building structure. In the event that copper or leaded copper sill pans are used, Subcontractor shall inspect each such pan to verify that all shop or field welds or soldered joints are properly made so as to insure the waterproof capability of the sill.
   C. All door sill pans shall have a vertical leg turn-up of not less than 1 inch.
   D. Every effort shall be made to avoid any penetration of a door sill pan. In the event that it is necessary to penetrate a door sill pan, the penetration shall be made waterproof. In this regard, special attention shall be paid to the installation of door thresholds to prevent penetration of the underlying sill pan.

3. In the event that solid vinyl windows are used, nailing of such windows shall be in strict accordance with the manufacturer's recommendations and special attention shall be given to verify that nails are placed in the oval slots provided by the manufacturer and not elsewhere in the window frames.

4. No solid vinyl window frame shall be racked, torqued, or twisted during installation.

5. No factory supplied window flashing shall be removed.

6. In the event that aluminum windows or doors are used, only stainless steel screws shall be used to install such windows or doors. No galvanized, black oxide coated, or other surface treated screws or nails shall be used.

7. In the event that Subcontractor has any question with regard to the requirements or recommendation of the Owner's waterproofing consultant, Subcontractor shall obtain the answers to such questions before proceeding with the work. In no event shall Subcontractor substitute Subcontractor's own judgment with regard to the means and methods of installation or choice of materials without prior written consent of Owner and Owner's waterproofing consultant.

Anti-Defect Provisions for the Framer

Addendum to contract with framer

This Addendum is made a part of and is incorporated into the contract between _____ Construction (General Contractor) and _____ , (Subcontractor).

In the performance of Subcontractor's work on the Project, Subcontractor shall do the following:

1. Subcontractor shall strictly comply with the plans and specifications and the requirements and recommendations of the structural engineer.

2. Unless otherwise specified by the plans, specifications, manufacturer's recommendations and requirements, and recommendations of Owner's consultants:

   A. Subcontractor shall personally instruct all framers with respect to the use of pneumatic nail guns and the installation of shear panels to insure that no nails are overdriven (overdriven nails are those that are set below the surface of the plywood or OSB shear panel). Subcontractor shall personally verify that the nail guns are properly set by observing the installation of nails in shear panels.

   B. Subcontractor shall verify that nails driven into shear panels are not more than 1/4 inch from the edge of same and that all nailing patterns as specified by the plans and specifications are strictly observed. Subcontractor shall personally verify that the correct type and length of nails are used to install shear panels.

3. All shear panels shall be numbered by Subcontractor on the plans. Subcontractor shall spray paint the corresponding number of the shear panel on the panel in the field so that Owner may verify that all shear panels called out on the plans are installed in the field.

4. Subcontractor agrees that Owner or Owner's representative have the right to inspect all shear panels for appropriate nailing and that if in the sole and exclusive discretion of Owner or Owner's representatives such panels are improperly nailed, Subcontractor shall replace the same with a new shear panel at Subcontractor's expense.

5. All hold-downs and anchor bolts shall be fastened securely at the time of installation and Subcontractor shall re-tighten the anchoring nuts on each of such assemblies prior to the installation of drywall.

6. Walls shall be plumb and in plane to the following tolerances: walls should be plumb within 1/8 inch for any 32 inches in vertical measurement. Walls shall be in plane without variance of more than 1/4 inch in every vertical 10 feet. (See, e.g., *Quality Standards for the Professional Remodeler*, 2nd. Ed., National Association of Home Builders Remodelors Council, Washington, D.C., 1991. Insurance/Warranty Documents, Home Owners Warranty Corporation, Arlington, VA, 1987; Spectext Section 06112, Framing and Sheathing, Construction Sciences Research Foundation, Baltimore, 1989.) Not less than two weeks after installation of the rough framing, walls shall be straight edged and planed as necessary to achieve such tolerances.

7. Subcontractor shall verify that all full height blocking has been installed to the correct dimension so that any roof sheathing assembly is in direct contact with such blocking.

8. Subcontractor shall verify that all horizontal fire blocking has been installed as per plans, specifications, and code.

9. No lumber with visible cracks, splits, wane, twist, or other defects that will affect the soundness of the installation or the plane or plumb of a wall, floor, or ceiling shall be used.

10. Metal hold-down devices shall be of a material that is compatible with and able to withstand, without corrosion, the chemicals used in pressure-treated lumber.

11. All plywood floor assemblies shall be "screwed and glued,"—i.e., installed using screws and construction adhesive to minimize flexing and noise.

Special Contract Provisions for the Plumbing Subcontractor

Addendum to contract for plumbing subcontractor

This Addendum is made a part of and is incorporated into the contract between _____ Construction (General Contractor) and _____, (Subcontractor).

In the performance of Subcontractor's work on the Project, Subcontractor shall do the following:

1. Subcontractor shall strictly comply with the plans and specifications and the requirements and recommendations of the Owner's representatives.

2. Unless otherwise specified by the plans, specifications, manufacturer's recommendations and requirements, and recommendations of Owner's consultants:
   A. Subcontractor shall personally instruct all plumbers with respect to the requirement to ream the cut ends of copper plumbing lines. Subcontractor shall personally verify by frequent inspections that all cut ends of copper plumbing lines have been reamed.
   B. Subcontractor agrees that Owner may make an unannounced site inspection and may remove portions of the copper plumbing lines to inspect for unreamed copper pipe ends. In the event that Owner determines through such inspection that pipe ends have not been reamed, Subcontractor agrees that the entire copper plumbing system installed to that time will be removed and reinstalled with properly reamed cut ends at Subcontractor's sole expense.
   C. All solder joints shall be free from excess flux and solder runs and shall have a clean, neat, and professional appearance.
   D. Subcontractor shall verify that all necessary dielectric fittings have been installed.
   E. All water heaters shall be properly strapped for seismic resistance and shall have PTR valves installed with drain pipes that are plumbed in a manner that water from such pipes does not enter any habitable living space.
   F. All venting shall be installed per applicable building codes.

3. Subcontractor shall strictly comply with all requirements and recommendations of the Owner's acoustical engineer with respect to the installation of supply lines and drain, waste, and vent lines and the isolation of the same from the framing of the structure. Subcontractor shall personally make frequent inspections to verify that compliance with such requirements has been achieved.

4. Subcontractor shall insure that no secondary drain holes in any deck drain or shower drain are plugged with debris or foreign material including mortar

5. Subcontractor shall take steps to protect the integrity of shower pan installations so that such installations are not harmed or penetrated by Subcontractor's employees after such installations have been completed.

6. Subcontractor shall instruct employees that no foreign material is to be placed in any drain or sewer line.

7. Subcontractor shall verify that all angle stops are properly installed, and that such angle stops for water closets are set at the appropriate height as called for in the Uniform Plumbing Code.

8. Subcontractor shall not penetrate the building exterior with any pipe or make any other penetration in the building exterior without prior written approval from the Owner's waterproofing consultant and without a specific recommendation with respect to waterproofing such penetration.

9. Subcontractor shall verify that all fixtures have an adequate supply of water and adequate water pressure and shall make such modifications or repairs to insure same at Subcontractor's expense.

Anti-Defect Provisions for the Stucco Subcontractor

Addendum to contract for stucco subcontractor

This Addendum is made a part of and is incorporated into the contract between _____ Construction (General Contractor) and _____, (Subcontractor).

In the performance of Subcontractor's work on the Project, Subcontractor shall do the following:

1. Subcontractor shall strictly comply with the plans and specifications and the requirements and recommendations of the Owner's representatives.

2. Unless otherwise specified by the plans, specifications, manufacturer's recommendations and requirements, and recommendations of Owner's consultants,
   A. Subcontractor shall personally instruct all lathers with respect to need to carefully install builder paper and lath on the structure. With respect to such installation, Subcontractor shall verify that
      1. Two layers of Type D builder paper (60 minute) or equivalent are used to wrap the building.
      2. There are no tears, splits, or other defects in the builder paper to allow water or moisture into the structure of the building.
      3. There are no reverse laps in any builder paper installation.
      4. No builder paper is installed over a horizontal surface.
      5. All overlaps of builder paper are adequate.
      6. All bib flashings are installed and tight to window and door openings.
      7. All fasteners used to adhere metal lath to the building are fastened to studs and not wall sheathing.
      8. All fasteners used to adhere metal lath to the building are embedded into the framing at least 7/8 inch.
      9. Metal lath protrudes from the surface of the builder paper an amount of space sufficient to allow the stucco scratch coat to key into the lath.

B. Subcontractor shall verify that no installer of builder paper or lath damages, removes, or destroys any flashing on any window or door opening or any flashing attached to any window or door.

C. With respect to the installation of the three coat stucco system, Subcontractor shall verify that

1. The scratch coat has been adequately keyed into the metal lath.
2. The scratch coat has been scratched sufficiently to allow the brown coat to key into the scratch coat.
3. The scratch coat is hydrated (watered) for at least 48 hours after installation.
4. All cracks in the scratch coat over 1/8 inch are patched before application of the brown coat.
5. The brown coat is hydrated (watered) for at least 48 hours after installation.
6. The proper mix of sand, Portland cement, and lime has been achieved with respect to all three coats of the stucco system.
7. The brown coat is smooth and flat in plane.
8. Any cracks in the brown coat are filled and patched prior to application of the finish coat.
9. All doors, windows, and other installations are adequately protected against damage from the installation the cementitious exterior plaster and that all excess plaster on such installations or on the ground around the building be removed promptly.
10. All weep screeds are installed and no weep screed is blocked by excess plaster in the holes in the weep screed.
11. No outside J box for an electrical fixture or installation of a plumbing fixture on the exterior of the building is covered over by stucco.
12. No scaffolding is permitted to be in direct contact with any overhead utility wire.
13. Penetrations through the builder paper caused by wire this used to secure scaffolding to the structure shall be sealed and made waterproof prior to application of stucco patch and finish coat.

ALTERNATIVE WAYS TO IMPLEMENT ANTI-DEFECT PROVISIONS

The best contract language in the world is useless unless the parties have a way to implement the provisions of the agreement. I prefer not to rely on the good faith of contractors and subcontractors, because even if they act with the best intentions, the construction process is such that many items can be overlooked or no record made with regard to their completion. Accordingly, I suggest that the best way to implement the special "anti-defect" contract language is to make a checklist of each of the items, for both the general contractor and the subcontractors, and to have the general contractor and subcontractor check off the individual items, date the inspection, and sign to verify that the item has been completed.

In this way, the owner can be assured that the items have in fact been done, and a written record can be maintained of when each item was verified and who verified it.

Like many of the issues discussed in this guide, it may seem very cumbersome, expensive, and time-consuming to create checklists and verify that responsible people have signed off on the various items. In reality, however, it takes little time and adds no expense to the project. A good builder who cares about the quality of the work will not object to the process.

7

Document Control

FOR CONTRACTORS

Construction projects not only create a mountain of dust and debris, they create a mountain of documents. Many of these documents are helpful to avoid construction defects, and many are helpful if there is litigation over a project.

In large-scale construction, formal protocols for document control are an important part of the process, and specific individuals are assigned to the task. In residential construction, informality is almost always the rule, and most contractors do not have additional personnel whose time can be devoted only to clerk-type tasks. Even so, some of the benefits of formality can be brought into residential construction and key documents must be maintained no matter how informal the construction process might be.

JOB LOGS

Construction disputes are almost always decided by looking backward. Months and sometimes years after the construction of the project, the owner, the contractors, the subcontractors, their experts, their attorneys, and the judge, jury, arbitrator, or mediator look backward and try to figure out what happened on a given day, a week, a month, or in the totality of a project. Without documents to show objectively what happened and when it happened, the resolution of construction disputes often depends on the memory and credibility of those involved, which is almost always a poor basis for reaching important decisions.

Job logs are a simple way of documenting the progress of a job and they can prove to be of tremendous importance in the event of a lawsuit. To many contractors, particularly in residential construction, job logs seem like a cumbersome and unnecessary task. One way to make the job easier is to create a form, have it printed, and put the form into a job notebook. This same notebook can contain other important documents (such as weekly meeting minutes, as described later in this chapter).

A job log form or daily report should contain the following information:

- Date
- Names of persons/subcontractors working on the job
- If contractor is performing self-employed work, the names of contractor's workers and their hourly rates
- Weather conditions
- Hours worked per employee and sub
- Scope of work (i.e., what was done that day and by whom)
- Name of person making report
- Shipments of goods and materials to the job site, by whom shipped, with bills of lading attached
- If materials are brought to the job site by a subcontractor, a list of the materials and a note whether the material was taken from the subcontractor's stock or purchased by the subcontractor for the project

WEEKLY MEETINGS

Weekly meetings are routine in large projects, but they can be of substantial benefit in smaller projects as well. Not only do weekly meetings keep owners up to date on the progress of the job, they also serve as a means for the contractor to inform the owner of potential problems, such as delays and cost increases. Recall that in the section dealing with communication between contractors and owners, delays and cost increases are a prime source of friction. Friction leads to conflict and conflict leads to lawsuits.

Weekly meetings should be documented in the form of minutes. Minutes can be prepared in a form, much like job logs. Weekly meeting minutes should include the following information:

- Date
- Time
- Location
- Names of attendees
- Topics discussed, both old and new business
- Decisions made
- Action (to-do) list

As a suggestion, contractors may wish to review a book of forms available from BNI Building News (www.bnibooks.com) entitled *Construction Project Log Book*. This publication contains forms including daily logs, contact information, checklists, and job meeting guidelines.

ESSENTIAL DOCUMENTS

The following documents should be considered essential to every job. These documents should be maintained in a file for at least 10 years following the earlier of the substantial completion of the project, the issuance of a certificate of occupancy, or actual occupancy. With regard to the time to keep records, in California, the statute of limitations for latent defects (defects that either cannot be observed or would not be understood as defects without expert advice) is 10 years from the earlier of the above three events. However, it is possible for subcontractors to be brought into defect litigation even after 10 years.

Here is an example. In 1995, a developer acquires a piece of property for the development of a tract of homes. The developer hires design professionals (architect, civil engineer, structural engineer, geotechnical engineer) to plan the project. The developer obtains a grading permit and mass grades the project. The economy changes and the developer decides to "land bank" the project until the market improves. Seven years later, the developer sells the property to another developer, who starts building. Seven years after that (now 14 years after the initial work on the project), a construction defect lawsuit is filed against the second developer, alleging that the soil compaction beneath one of the homes is defective and the home is breaking in half. The second developer sues the first developer, under whose "watch" the soil was graded and compacted. The first developer sues the grading contractor and the design professionals. Even though it has been more than 10 years since the grader and the design professionals started their work, they would not be able to extricate themselves from the lawsuit on a statute of limitations basis, because the law, at least in California, provides that they can be brought into the lawsuit on a claim for indemnity.

The following documents should be maintained:

- The owner/contractor contract, with all changes and addenda
- All contracts between the general contractor and the subcontractors
- All change orders
- All requests for information (RFIs) and responses thereto
- All shop drawings
- All submittals
- All policies of insurance for the general contractor and the subcontractors
- All additional insured endorsements, under the terms of which the general contractor or the owner was named as an additional insured under the contractor's policy or any policy of the subcontractors
- The plans
- The specifications, if separate from the plans
- The written recommendations of owner's consultants
- Correspondence between the owner, contractor, and subcontractors and consultants
- Invoices

- Receipts
- Meeting memos
- Field directives
- Daily reports (job logs)

INSURANCE POLICIES

There is a difference, and the difference is a critical one, among the entire insurance policy, a certificate of insurance, and an additional insured endorsement. A *certificate of insurance* confirms that a contractor or a subcontractor has insurance, it identifies the name of the insurance carrier, and it states the amount of the insurance. An *additional insured endorsement* merely confirms that a certain person, usually the contractor or owner, has been named as an additional insured under the policy. But neither the certificate of insurance nor the additional insured endorsement give any real information about the policy itself.

The *policy of insurance* contains all of the important details, including the type of insurance and policy exclusions and endorsements. Without knowing the type of policy and the policy exclusions, the holder of the certificate or additional insured endorsement might have a policy that is virtually worthless and not know it. For example, the contractor or subcontractor might have a "condominium exclusion" in the policy. If the work in question is a condo project, the owner might think that he or she is dealing with an insured contractor or subcontractor, but in reality, there is no insurance available because of the nature of the exclusion.

There is another important reason why it is essential for owners to have the contractor's complete policy of insurance and for contractors to have the subcontractor's complete policy of insurance. Because the statute of limitations for construction defects is so long (at least 10 years for latent defects under California law), and because it is not at all uncommon for contractors and subcontractors to go out of business, it is entirely possible that 10 years after the building is completed and a lawsuit is filed, the contractor, subcontractor, and all of their files are long gone. The owner now wants to sue the contractor. The contractor, having been sued, wants to sue the subcontractors. But the contractor is gone, and the subcontractor is gone. The only party left who can respond is the insurance carrier.

Without the policy, there is no way to contact the insurance carrier and, many times, no way to prove to the carrier that the contractor or subcontractor was insured by them. Insurance carriers purge their files, and policies and records are sometimes lost or destroyed. A copy of the entire policy will end the discussion and will force the insurance carrier to respond to the lawsuit.

8 Managing Conflict During Construction

FOR OWNERS AND CONTRACTORS

Despite the best efforts of all concerned, it is inevitable that some conflict will arise during construction. The real question is how to manage that conflict so that the conflict does not result in the job being shut down and a lawsuit being filed. At the outset, good communication between the parties will go a long way toward avoiding and resolving conflict, and the techniques described in Chapter 3 should be employed first.

In the event that good communication does not resolve the conflict, several alternatives are available for managing conflict and achieving resolution of disputes without disrupting the flow of construction. Some of the alternatives are more practical for larger projects, but to some extent, all of the concepts can be applied to smaller projects as well.

MEDIATOR

An approach that is used increasingly is the concept of a *course of construction mediation*. In this approach, before the project starts, the parties (usually just the owner and contractor, but sometimes the prime subcontractors) agree on a neutral third-party mediator. This mediator is empowered to review the contracts, visit the site, review the plans, and otherwise familiarize himself or herself with the project and keep up to date on the progress of the construction.

If a dispute arises, the parties immediately turn to the mediator for assistance in resolving the conflict. There are some singular advantages to this approach. The mediator is already familiar with the project and the parties. The parties already trust the mediator and have confidence in the mediator's ability to help them resolve the conflict. The process of conflict resolution can start immediately: there is no delay while the parties search for a mediator and bring the mediator up to speed. Finally, the mediator will have sufficient construction experience and expertise to assist the parties effectively. In other words, the mediator is *pre-qualified*.

MEDIATOR/ARBITRATOR

A second approach is *course of construction mediation/arbitration*. In this approach, a mediator is selected, as described. The parties participate in mediation, as described. However, if the mediator is unable to help the parties reach an agreement, the parties empower the mediator to "switch hats" to that of an arbitrator, and the parties let the mediator/arbitrator decide the issue in a binding decision.

Although many parties are reluctant to provide the mediator/arbitrator with this authority, one thing to remember is that if the parties do not resolve the issue by themselves (or with the mediator's assistance), the dispute will ultimately be submitted to a third party, whether arbitrator, judge, or jury. That third party will decide the issue for the parties.

If the mediator/arbitrator concept is put into place, at least the parties have the comfort of knowing that the mediator/arbitrator has construction experience, the mediator/arbitrator understands the issues and is familiar with all the background, and the parties all respect the experience and wisdom of the mediator/arbitrator.

PROJECT NEUTRALS

A third approach is the use of *project neutrals*. In this approach, a panel of neutral mediators/arbitrators is selected by the parties to advise them with respect to dispute. This panel can hold various degrees of responsibility, up to and including the right to arbitrate dispute and render a binding award.

9

Cost-Saving Techniques

FOR OWNERS

Sticker shock is a fact of life in the construction business. Homeowners are invariably surprised by the cost of labor and materials and invariably worried that the cost of construction will increase between the start of the job and the day that the job is completed. However, there are ways to save money, or at least to control costs, in the construction process. Cost savings in construction requires patience, planning, and perseverance.

Cost saving in construction results from the following major categories: (1) planning before construction starts; (2) shopping for materials before materials need to be installed; (3) minimizing changes during the construction process; and (4) general contracting the job yourself.

PLANNING

Planning enables the owners to accomplish several important things to save money. Planning allows the owner to choose materials well before the time that the materials are due on the job site. This time delay, in turn, allows the owner to shop intelligently for materials, to take advantage of sale prices and special discounts, and to obtain materials at the lowest price possible.

Planning minimizes the need for change orders during construction. Changes during construction add significantly to the overall cost of a project. Many times, contractors charge on a time and materials basis for changes. These charges can be very expensive.

What kinds of decisions need to be made during the planning stage? The answer lies in understanding where the costs of construction can be controlled and where costs are more or less fixed. The "sticks and bricks" aspects of construction, meaning the cost for materials for concrete, framing lumber, rough plumbing materials, insulation, stucco, and roofing, are not subject to significant differences at the consumer level. Large developers who build thousands of units are able to achieve significant cost savings due to the large volume of materials purchased, but these benefits are not generally available to consumers.

There are at least two exceptions to the preceding statement. The first is framing lumber. It is possible to save money on framing lumber. Many framers will enter into a labor-only contract, with the owner supplying materials. The framer will generally supply the owner with a list of the materials needed, which the owner can use to shop prices. Many times suppliers in outlying areas have better prices and are willing to truck the materials to the jobsite. It is possible, with enough research and bargaining, to save 15 percent or more on the cost of framing lumber. There is, however, a downside to the process of hiring a framing contractor on a labor-only basis—the contractor has no incentive to use the lumber efficiently because the framing contractor did not buy the lumber. It has been my experience that the waste involved may outweigh or equal the initial cost savings.

The second, and more important exception, is windows and doors. A huge variety of windows and doors are on the market, and there is considerable variance in the expense for these items, even among items of seemingly the same or roughly similar quality.

As a practical matter, with the caveat of windows and doors, the greatest expense, and the greatest potential cost savings in any residential construction project, are fit and finish items. These include plumbing fixtures (bathtubs, lavatory sets [faucets], sinks, toilets, shower heads, shower controls, washer/dryer), electrical fixtures (recessed lights, receptacles, switches, lighting control systems, exhaust fans), wall coverings, floor coverings, moldings, cabinets, countertops, closet fixtures, garage fixtures, laundry room fixtures, and appliances.

PURCHASING MATERIALS

There are many ways to save money in purchasing materials. The first, and most effective, is to keep the choice of materials consistent with the overall level of quality in the project. In other words, and to put the matter in blunt terms: do not buy "Rolls Royce quality" materials for a "Ford project." There is a tremendous temptation in construction to want something just a little (or sometimes, a lot) better. This temptation is rationalized in many ways, the most common being the "While we're at it" or "We might as well" syndromes. But most of the time, this extra expense (which can add up rapidly) is simply not worth it. For example, in plumbing fixtures, many of the larger manufacturers now make products at reasonable costs that in many ways are identical to the highest-end products, without a significant loss of quality. Because plumbing products wear out, and because styles change, it is rarely a good investment to purchase the highest price products when a comparable product at less cost might look virtually identical and perform just as well.

You can save substantial amounts of money by smart shopping. The growth of online retailing has reduced prices enormously for many household items, including plumbing and electrical fixtures. Many e-retailers have small overhead and are able to pass the cost savings along to retail customers. Even eBay lists tens of thousands of electrical and plumbing fixtures.

Here are a few examples from my own experience. In the course of building a new home in 2003 and 2004, we purchased a large number of items by Internet. Our stainless steel drawer pulls averaged $6 to $10 each. Retail, these would have cost at least three or four times that amount. Our bathroom faucets (mostly Italian contemporary design) were, on average, about $100 each. Comparable faucets retail were $400. Our glass vessel sinks averaged $110. Comparable sinks retail were five times that amount.

In any large urban area, the large plumbing supply houses will usually have discount outlets where they sell discontinued items, overstocked items, returns, or floor samples. Diligent shopping at these outlets can lead to some astonishing results. Our floor sample spa tub was 15 percent of the cost of a new tub, in the box and in perfect condition.

In addition, many of the large designer type showrooms have yearly sales. For example, at the Los Angeles headquarters for Ann Sachs, a nationally known distributor of fine tile products, a large warehouse in an industrial section of the San Fernando Valley holds a yearly parking lot sale. Many odd lots, discontinued items, out of style items, and trim pieces are sold for a fraction of their original retail cost. In the course of building our own home, we purchased limestone from this parking lot sale for about $2.50 per square foot: the retail price would have been at least $10.00 to $15.00 per square foot.

Another excellent source of building materials is local recycler-type newspapers. Each of these newspapers has a building materials section that lists hundreds of new and used products. Materials that are uniformly available through these advertising publications include windows, doors, lumber, brick, tile, roofing, plumbing fixtures, electrical fixtures, and more. Many times these materials are brand new and can be purchased for pennies on the dollar of their original cost. In our own home, our large spa tub for our master bath was purchased through the *Recycler*, a local paper. We bought the tub from a plumbing contractor who had ordered one extra for a housing development. Our price for this $4500 tub was $1700, new, in the box.

MINIMIZING CHANGE ORDERS

The obverse to the planning coin is the concept of minimizing change orders after construction begins. If sufficient planning has occurred, very few change orders should be necessary, and they should be limited to those caused by unexpected site conditions.

As a practical matter, change orders usually result from the "While we're at it" syndrome. As the building progresses, and the owner can actually see the physical dimensions of the structure (and not just lines on a piece of paper), the desire to make changes becomes almost irresistible. Phrases like, "Let's make this bathroom just a little bit bigger," or "If we move that wall, we can have more space," become common.

What's hard to remember is that there is a cost associated with moving a wall, relocating an electrical outlet, or moving the location of a plumbing fixture. It is likely that all such work will be done on a time and materials basis, because the subcontractor involved cannot mobilize an entire crew to make small changes, and economy of scale is lost. There is also a ripple effect, because the changes slow down the progress of the job and impede the ability of other subcontractors to complete their work in the areas where changes have been made.

Finally, it is rare for a change order to reduce and not increase the cost of construction. Most changes involve the use of better materials, not lesser materials. They involve building more, not less. They end up creating delay, not earlier completion. All of these items have an impact on the bottom line.

OWNER/CONTRACTOR

Acting as your own general contractor has at least the greatest potential cost savings in any building project, because a general contractor will customarily charge 20 percent of the overall job cost for his or her overhead and profit. By eliminating the general contractor, you automatically save this 20 percent. However, this course of action is fraught with considerable peril.

The best place to start on a discussion of general contracting your own project is to review what a general contractor actually does and what you are paying for when you hire a general contractor. At the most basic level, a general contractor does the following:

- Provides labor, either from his/her own employees or through subcontractors
- Provides materials by finding and ordering same
- Schedules the work of his/her own employees or subcontractors
- Coordinates the work of his/her own employees or subcontractors
- Supervises the work to ensure quality control
- Handles accounting and lien releases

The actual work done by a general contractor varies considerably. Some general contractors have employees who perform many, sometimes all, of the various "trades" that are usually the province of the subcontractors. This type of general contractor schedules all the labor using in-house personnel and coordinates the job using what is called *self-performed labor*. Some general contractors provide some skilled, some semi-skilled, and some unskilled labor. This type of general contractor uses some in-house labor and much subcontractor labor. Some general contractors provide no labor whatsoever.

A review of the preceding list discloses that an owner/contractor must assume responsibility for the same six basic categories of work. What does this mean in the real world, and is it possible to do without a general contractor and still obtain a quality result?

First, there is the question of providing labor. Since you, as the owner, do not have employees, you must locate qualified subcontractors. Because every job requires at least some unskilled and some semi-skilled labor, you must locate these individuals as well. Unskilled laborers perform such tasks as cleaning up, moving materials, and digging. Semi-skilled laborers assist the subcontractors and can perform such tasks as constructing basic formwork for the pouring of concrete or demolishing existing structures.

Locating qualified subcontractors is not an easy task and is the single biggest challenge that you will face as an owner/contractor. There is tremendous variance in skill, price, reliability, honesty, integrity, and competence. In most areas, the best subcontractors work repeatedly and consistently with the best contractors. They are reluctant to work directly for owners for two reasons: because they perceive owners as amateurs who will create chaos, and because they obtain repeated work from good contractors. So they do not wish to have a single owner/contractor job interfere with their availability should, one of their good contractor customers call them to work on a project.

Second, you will have to provide at least some of the materials for the job. You will have to locate building supply houses and obtain trade discounts from those suppliers. This is a time-consuming endeavor. The better building supply houses are not the average big box store. They are typically located in industrial areas and can be difficult to find. Many times they will not extend trade discounts to owners who are not licensed contractors.

Transporting materials can also be difficult. Although many of the larger supply houses will deliver, it seems that there is always the need for a pickup truck full of some material. If you happen to own a pickup truck (or something larger, which is also frequently necessary), your work will be a little easier. If not, renting a truck repeatedly can become an expensive burden.

Third, scheduling the work requires an enormous commitment of time and energy. Getting people to a construction job, and getting them there on the appointed day at the appointed time, is a Herculean task. The construction industry has a bad reputation (some people think it is well-deserved) for reliability. Subcontractors, for a variety of reasons, are notorious for not showing up on time and will offer many wonderfully creative excuses. You will have to commit yourself (perhaps by the end of the job you will be ready to commit yourself) to a full time job on scheduling issues alone.

Fourth, the job must be coordinated. There is a slight, but important difference between scheduling and coordination. Coordination requires an understanding of sequencing— that is, what aspects of the job must proceed in what order. Additionally, coordination requires an understanding of what trades need to be on the job at the same time, or different times, and how many trades can be on the job at once without causing disruption and inefficiency.

Next is the issue of quality control. At the outset of this book I made the observation that nothing in these hundreds of pages can substitute for the years of experience of a good contractor. In some respects, quality control is the single most important aspect of the contractor's work. A project can proceed beautifully, on time and on budget, but if the result is poor due to faulty workmanship and bad quality control, the fact that it was finished on time and on budget is meaningless.

It is probably impossible for the average homeowner to obtain good quality control without outside help, simply because the average homeowner does not have the years of experience to see all the problems and issues. There is, however, a solution for this that does not require the services of a general contractor: hire a qualified job superintendent or an owner's representative/construction manager to oversee the project from a quality control perspective.

In recent years, the industry has seen increased use of construction managers and private job superintendents. Construction managers act as the owner's representative. They are the owner's eyes and ears on the project and they answer only to the owner. They are not there to make friends with the subcontractors or to keep the subcontractors employed from job to job. A qualified construction manager will observe the progress of the job and report to the owner on problems, including construction issues, scheduling issues, and potential cost overruns or delays. In some cases, the construction manager will agree to give instructions directly to the subcontractors and in this regard he or she will function more like a job superintendent.

A private job superintendent performs a similar function. A job superintendent, unlike a construction manager, is dedicated to one single project and will be on the jobsite every day during the project. The job superintendent will have authority to direct the work of the subcontractors and can assist in scheduling and coordinating.

How do you choose a job superintendent or construction manager? Construction management firms are not difficult to find. They advertise, and some of them are large national or even international companies. Not all of these large organizations will provide construction management services for single family homes, but they are usually a good source of information about smaller companies that will provide such services. Job superintendents are a different matter. My own philosophy on job superintendents is that the perfect superintendent is a person who is newly retired from a construction firm that specializes in building homes. These individuals usually have 30 years or more experience, and they are usually interested in the possibility of earning a good salary on a single project.

The cost of a construction manager or job superintendent does add to the overall expense of the project. Construction managers charge by the hour. Their charges, for principals with the most experience, are in the range of $150 to $185 per hour. Keep in mind that the construction manager will not be on the job every day all day. Other, somewhat less experienced personnel from the construction manager's office will charge less, usually in the $75 per hour range. A good job superintendent will charge in the range of $800 to $1000 per week. For the person with the right qualifications and the right temperament, this could be money well spent.

10

The Lawsuit—What Happens Next

FOR OWNERS AND CONTRACTORS

Despite every effort to avoid and resolve conflict, sometimes disputes end up in litigation. At that point, you need to know the best way to deal with the lawsuit. To begin, you need some background on the process of construction defect litigation and how to deal with it.

Construction lawsuits are determined in one of the two ways: in a court of law, with either a judge or a jury hearing evidence and rendering a verdict or judgment, or in an arbitration, where a trained and experienced arbitrator hears evidence and renders an award that binds the parties and has little review by a court.

As a matter of practicality, however, nearly all (in the range of 99 percent) of all construction disputes never make their way into either a trial or an arbitration. Nearly all disputes are settled in a mediation environment.

THE MEDIATION PROCESS

Mediation is a process where a trained and experienced neutral third party helps the disputing parties reach a resolution. Unlike a trial in court or an arbitration, a mediator does not decide the issues. The mediator does not decide who is "right" and who is "wrong." The mediator does not decide who wins and who loses, or how much money is to be awarded.

In construction matters, particularly, the mediator does two things: he or she helps the parties understand their risk, and he or she helps the parties manage that risk. What makes a mediator effective in accomplishing these tasks and how do you find an effective mediator? The first step in finding an effective mediator is understanding what makes a mediator effective. The second step is determining whether your proposed mediator has that set of qualifications.

Effective mediators have a skill set that includes, to varying degrees, patience, persistence, insight, communication skills, professional training, experience, wisdom, technical expertise, and a passion for the work. A sense of humor doesn't hurt.

No one skill, or even one set of skills, necessarily makes a mediator effective. For example, some lawyers believe that retired judges make effective mediators because they have substantial experience in handling settlement conferences, and retired judges have decided (i.e., resolved) many disputes during their years on the bench. It is not necessarily true that every retired judge is an effective mediator, although many of the best mediators are in fact retired judges. Retired judges often have no formal mediation training. They often are more accustomed to being listened to than listening to others. Their training has been in what mediators call "distributive bargaining," which is simply moving money from one side to another, usually without much regard to the merits of the dispute. Retired judges usually conduct what are called "evaluative" mediations, in which the mediator evaluates the merits of each argument and gives the parties an opinion about the relative merits. Not every mediation benefits from this approach. In many instances, a "facilitative" approach is more effective, and many retired judges do not have training to apply this technique.

Surprisingly, perhaps, although most lawyers assume that a formal legal background would be an absolute threshold requirement, some of the best mediators have no formal legal training. These individuals may bring to the mediation practice a business background, training in psychology, or communication techniques that make them extraordinarily effective in resolving disputes, even disputes that involve complex legal and technical issues.

Many mediators specialize in a particular subject matter area, such as construction and construction defects. They develop, through repeated interaction with the subject matter and with the lawyers and insurance claims representatives, a familiarity with the typical areas of dispute and the "hot-button" issues. This base of knowledge of the technical aspects of the disputes, and their familiarity with the "players," enables them to resolve disputes quickly and easily.

Generally speaking, mediation services in most large cities are provided by three groups or categories of service providers: Large dispute resolution services that use a mix of retired judges and lawyers (most work full time as mediators and some maintain their own practices); the American Arbitration Association, which uses part-time and full-time mediators, usually also a mix of retired judges, lawyers, and nonlawyers; and private mediators, generally lawyers who practice mediation full time.

A large mediation service will provide upon request a roster of mediators with a description of the mediator's particular background and expertise. Experience has shown that the best way to determine the most effective mediator from the list is to call the administrator of the service provider and discuss the choices. You should be prepared to discuss in general the nature of the dispute, any particular personality issues that you believe might be important in the dispute, and technical expertise that would be helpful. The administrator should be able to recommend a particular mediator who might be well suited to help resolve the dispute. Mediators tend to acquire a reputation in the community quickly. As a result, references from other lawyers are an invaluable source of information about a potential mediator. You can call these lawyers and gain an insight from them as to the mediator's style and expertise.

The private mediators are uniformly ready and willing to discuss their qualifications and experience in an informal way, so that you can assess them before formally retaining their services. This informal discussion is an excellent way in which to gain insight into the proposed mediator's character and approach. During the conversation, ask yourself the following: Does the mediator listen well to your comments? Is the mediator patient during the conversation? Does the mediator have the technical background to understand the issues? Does the mediator seem highly motivated to help you resolve your dispute?

If you believe that you need further information, on request, most mediators will meet with you before being retained to discuss in more detail their qualifications, experience, success rate, approach, and methodology.

After you have selected your mediator, you must use the mediator effectively to resolve your dispute. First, you need to prepare your case. This may sound simplistic, but, astonishingly, many lawyers and clients walk into mediation expecting that because the process is so informal, "winging it" won't hurt their case. Unfortunately, that's not true. In a successful mediation, information is everything. That means that the successful lawyer and the client will have the necessary information to demonstrate the validity of the claim or the defense to the claim, and they will have that information in a form that can be readily and convincingly communicated to the other side. Clients need to prepare for mediation as well. Preparation means that the client must enter into the mediation process with a commitment to settle the dispute and with an understanding that making the mediation simply another round of the adversary setting is not productive.

The decision-maker must be present, and sometimes the decision-maker isn't the person that you would expect. Here's an example: A dispute occurs between a homeowner and a construction company's sales representative. You might think that the head of the construction company is the decision maker, but some issues require the salesperson's involvement to resolve the dispute. The homeowner might, for example, want a personal apology from the salesperson before a settlement can be reached. In such a situation, the salesperson's participation in the mediation might be just as critical to a resolution as that of the owner of the construction company. Another example is a dispute between a homeowner's association and a homeowner, where the entire board must be present to approve the settlement. In those situations, it is not uncommon for the board to send one or two members, in which case the settlement cannot be concluded because the board is not present to vote.

Next, you need to communicate honestly and openly with the mediator about the good and the bad aspects of your position. The mediator will assure you that anything said in a private session will not be disclosed to the other side without your permission, and every good mediator religiously honors that commitment. Many lawyers hear that commitment and immediately disregard it. They then proceed with the mediation process by telling the mediator that they will pay, or accept in payment, a particular number, when in reality they will pay substantially more or accept substantially less. This approach is not productive and it delays the resolution process. Remember that there is a difference between telling the mediator what you would pay, or what you would accept, and reaching that number. You and the mediator will devise a strategy to obtain that settlement, but that strategy will be more effective and take less time if the mediator learns your true goals in the mediation.

Remember that the intention is to reach a resolution of the dispute. Reaching a resolution requires both parties to focus less on the merits of their position and more on the benefits of resolution. (Remember that no case is ever perfect, and it is not only difficult to determine with certainty who is right or wrong, but the concepts of right and wrong often aren't the determining factor when lawsuits end up in front of a judge or jury.) To the client, the benefits of resolution are many, and they include the fact that most cases will eventually settle without a trial. An early settlement will save the client not only tens of thousands of dollars for attorney's fees and costs, but will also reduce the lost income that results from time away from his or her business affairs. It will also bring certainty to the litigation process and allow the client to make a business decision that the client controls.

Give the process time to work. Mediation is a process, and like every other process, it takes a certain amount of time to do it right. In complex disputes, for example, it is not unusual to have multiple sessions of mediation. While at first blush this might seem to be a large commitment of time, consider that if the dispute were to be resolved in a lawsuit, scores of days would be spent in depositions and in preparing for trial, not to mention days spent in court.

Have confidence in the process. Statistically, an effective mediator will have a high settlement rate. The odds are therefore much in your favor that the mediator will help resolve the dispute and resolve it to the client's satisfaction.

Doing your homework before the mediation process by knowing what skills are required of an effective mediator, choosing the mediator carefully, and preparing for the mediation will make your mediation faster and more likely to resolve your dispute.

THE ARBITRATION PROCESS

What happens if you don't resolve your case at a mediation? In most cases, construction matters are then decided in arbitration, because most construction contracts call for arbitration as the forum in which to decide the dispute.

Arbitration is a process whereby an experienced professional person, the arbitrator, instead of a judge or jury, listens to evidence and arguments, and decides the dispute on the merits, by issuing an award in favor of the winning party. Although arbitration is somewhat like a trial in a court of law, arbitration is more informal than a trial in court, and arbitrators have more discretion to be fair than judges, who must follow strict legal rules. Arbitration is also more cost-effective than court proceedings because arbitrations generally take less time than lawsuits in court. The decision of the arbitrator is final and generally no appeal is possible.

Arbitration takes place when both parties agree in advance as part of their contract (the parties can always agree to arbitrate after the contract is signed if a dispute arises later). The parties decide for themselves what kind of arbitration they want and what kind of arbitrator they want to hear the dispute. For example, the parties could agree to place time limits on the length of the arbitration, to limit the damages, or to have a person with substantial construction experience act as the arbitrator.

Arbitrations are generally conducted by private dispute resolution services, the largest by far being the American Arbitration Association (AAA). Other similar organizations are Judicial Arbitration and Mediation Service (JAMS), ADR Services Inc. and the Forum Dispute Management. All of these organizations have established rules and procedures and provide trained and experienced arbitrators and administrators to handle the case from start to finish. The AAA and similar arbitration service providers charge a fee for administering the case, and the arbitrator charges a fee for the hours spent hearing and deciding the case, usually on an hourly basis. In their contract, the parties can agree on which dispute resolution service to use and what kind of arbitrator they wish to hear the case—for example, a lawyer with at least 20 years of experience in construction, a general contractor, an architect, or a retired judge.

The administrative fee is a percentage of the amount that the claimant is seeking. The arbitrator's fees vary, but they are determined by the arbitrator's hourly rate. The hourly rate for experienced arbitrators is usually $350 per hour or more and is generally shared equally by the disputing parties. If the contract provides for it, the arbitrator can award the winning party all or part of the administrative costs of the arbitration and attorneys fees as well. Here's a tip for both owners and contractors: if your contract contains an attorneys fees provision, the arbitrator has the right to award attorneys fees to the "prevailing party." But be careful—sometimes the attorneys fees become so substantial that they end up creating a bigger problem than the underlying dispute. Often, although it's not the conventional wisdom, not having an attorneys fees clause discourages litigation. Here's another tip: Construction disputes almost always involve expert witnesses, and experts are expensive. You can provide in your contract that the winning party is entitled to recover the cost of its expert witnesses.

Starting the Arbitration Process

The arbitration starts with a *demand for arbitration* from the person who is seeking to be paid, accompanied by a *claim* setting forth the amount that the *claimant* (the person filing the claim) is demanding. The demand for arbitration is sent to the person from whom payment is sought (the *respondent*). If a homeowner sues a contractor, the homeowner is the claimant. If the contractor believes that the homeowner owes money, the contractor could file a counter-claim and that counter-claim would be heard in the same arbitration as the homeowner's claim.

The arbitration service provider will give each party a list of proposed arbitrators. Each party can disapprove a certain number of arbitrators until an arbitrator is selected, either by mutual agreement or by the provider.

The Arbitration Hearing

Arbitration is similar to a trial, but more informal. The parties are usually represented by lawyers. The arbitration takes place at the arbitration service provider, at the offices of one of the lawyers, or at a place agreed to by the parties (sometimes at the site of the construction dispute). Before the arbitration hearing starts, the parties will have had a pre-hearing conference with the arbitrator to discuss the format of the arbitration and any issues that should be resolved before the arbitration hearing starts. The parties will also have exchanged the documents (called *exhibits*) that support their claims, and they will have given the arbitrator legal briefs explaining their positions. They will also have disclosed the names of any witnesses.

The arbitration hearing starts with the lawyers or the parties (if there are no lawyers) giving a short statement of their positions. Then the claimant calls witnesses to testify in support of his or her position. The witnesses are sworn by the arbitrator and testify under oath. Witnesses can be subpoenaed to testify at the arbitration. The respondent is entitled to cross-examine the witnesses and the arbitrator is also entitled to ask questions. Depending on the length of the proceedings, the arbitration could start and finish on consecutive days or stretch out the hearing over several weeks with sessions held every few days. The arbitration schedule often changes to accommodate the schedules of witnesses, and arbitrators have broad discretion to allow continuances so that all parties have the chance to present their evidence.

Typically the witnesses in a construction case would be the owner, the contractor, subcontractors, design professionals, and expert witnesses.

The arbitrator listens to the testimony, reads the exhibits, and listens to arguments from both sides. When all the evidences have been presented, the arbitrator will either close the hearing or may ask the parties to submit additional legal briefs. Once the hearing has been closed, the arbitrator must issue an award within 30 days. The award is in writing and may be as simple as "The Arbitrator awards the claimant the sum of $X.00," or it may contain the arbitrator's reasoning, with references to the evidence.

In rendering a decision, the arbitrator is not required to follow the strict rule of law (unless the contract so states), but instead is entitled to render an award that is basically fair to the parties. This is one of the biggest single differences between arbitration and litigation. In court, the judge and jury are required to follow strict legal rules. This is an important consideration when deciding whether to arbitrate a dispute. If you believe that fundamental fairness is on your side, you will probably do better in arbitration than in court. The other principal difference is that the rules of evidence do not apply in arbitration unless the parties agree otherwise. This means that hearsay and other forms of evidence that could not be heard in court are perfectly permissible in arbitration, with the arbitrator deciding what weight to give to the evidence.

Length and Costs of an Arbitration Hearing

Arbitrations can be short and simple—less than a day—or they can last for many weeks, depending on the complexity of the case. In most instances, arbitrations take less time than a trial in court because the informality of the proceedings speeds up the process. Usually the hearing will take place more quickly than if the parties had filed in court. Most arbitrations are completed in nine months or less from the time the demand for arbitration is filed to the time the award is issued. It is difficult to generalize about the cost of an arbitration, but between the arbitrator's fees and the legal fees, it can cost several thousand dollars per day. One reason that arbitration is less expensive than a trial in court is that arbitrations do not include formal discovery. *Discovery* is a process where the parties take depositions, send written questions called *interrogatories*, exchange demands to produce documents, and otherwise learn about the case before the trial. While discovery is often helpful for each party to learn about the claims and defenses in the case, it is also expensive. In arbitration no discovery is permitted unless the parties agree otherwise, either in their contract or during the time before the hearing starts.

Establishing Awards and Paying the Lawyer

The arbitrator cannot award more than the amount claimed, plus costs and attorneys fees if allowed by the contract between the parties. The lawyers are paid in the same way that they would be paid in any lawsuit, either on an hourly basis or on a contingency basis. Although it is unusual, the parties can agree in their contract to limit the amount that the arbitrator can award. Like all legal proceedings, arbitration is not inexpensive, but it is almost always less expensive than going to court.

After the award is issued, either party can ask the court to *confirm* the award, which has the effect of making the award a court judgment. If the other party objects to the award, that party might ask the court to *vacate* the award. The decision of an arbitrator can be vacated for only two reasons: the arbitrator failed to disclose a conflict of interest that is important enough that the parties would not have used the arbitrator to decide the case if they had known of the conflict, and/or the arbitrator failed to allow the parties to present evidence in the arbitration. Otherwise, the arbitrator's decision is final. The winning party can proceed to collect on the judgment as if it were issued by a court or jury. This does not mean that the winner collects automatically; it means that the winner will have a judgment and can enforce that judgment using the legal process.

Developing a Strategy to Win at Arbitration

Many articles have been written about strategies for successful arbitration, but it really comes down to a few important issues. Arbitrators are concerned with and impressed by professionalism, reasonableness, and fairness. This means that if you and your lawyer are well prepared and well organized, you will do well. As an arbitrator, I am not impressed with lawyers and parties who argue with one another over trivial issues or who try to take advantage of one another. Arbitrators are interested more in fairness than in legal technicalities. If you and your lawyer concentrate more on what is fair and less on technical legal issues, you will do well. Arbitrators are always interested in the truth. If you and your lawyer are truthful and forthright, if you admit what must be admitted (even if you think that it is "bad" for your case), and you demonstrate that the other side has not been honest, you will do well. As an arbitrator, I am immediately suspected about a party's entire case if I believe that the party or their lawyer has lied about even a small aspect of the overall case. Arbitrators can most easily understand a case that sticks to the facts and a story that is logical and to the point. If you can make your case brief, focus on the important issues, and leave out what is not essential to proving your point, you will do well.

A Lighthearted Look at Arbitration

Sometimes the best way to learn how to do something right is to know how it's done wrong. Here's a lighthearted view of the arbitration process, entitled "10 Foolproof Ways to Lose Big at Arbitration." This article was written by me and published by the *Los Angeles Daily Journal*, California's largest legal newspaper, in 2004. It is reprinted with permission. Although intended primarily for lawyers, many of the concepts are important for both owners and contractors.

10 Foolproof Ways to Lose Big at Arbitration

Endless articles have been written about how to get to arbitration and what to do after you arrive to present a persuasive and winning case. Strangely, very few articles are devoted to the best ways to lose at arbitration. A clear guide, written from an insider's perspective, is obviously long overdue.

1. Whatever you do, don't read the rules.

   Arbitration is supposed to be informal, right? So why should you bother with the rules of the alternative dispute provider? They can't be all that important. Be sure to avoid any consideration of the rules pertaining to the selection of the arbitrator, so that you end up with someone with little or no experience in your dispute. Also, you will definitely want to avoid knowing the rule that says that you can expand or contract the way in which the arbitrator will handle your case and whether the arbitrator will have to write a "reasoned opinion." After all, who would want to know how the arbitrator decided the case—don't they all just split the baby anyway?

2. While you're at it, don't read the contract either.

   So what if you have contractual arbitration and the contract defines the scope of the arbitration. You'll get to the contract sooner or later in the arbitration. Why should you read it before-hand? Besides, if you let the arbitrator know the limits of his or her authority during the arbitration, and not before, it will be fresh in the arbitrator's mind. Anyway, explaining all of these details, such as whether attorney's fees are recoverable, whether the arbitrator has to follow the rule of law or rules of evidence and whether any discovery is available is a lot of trouble. The arbitrator does this all the time, let him figure it out.

3. Arbitrators are all the same, why bother to check them out?

   Everybody knows that all arbitrators are the same. Why waste your time in reading those lengthy arbitrator resumes and calling around to other lawyers or friends in the community to find out whether that arbitrator is smart, fair and reliable? Aren't all arbitrators fair, experienced, reasonable, patient, careful and conscientious? They must be, or they wouldn't be arbitrators.

4. Be an unreasonable jerk during pre-arbitration hearings to show how tough you are.

Those pre-arbitration hearings aren't just for scheduling and resolving potential procedural issues to streamline the case, they're a great opportunity to show you are a tough-as-nails, take-no-prisoners litigator, a stance sure to impress any arbitrator. You could start by refusing to stipulate to facts that everyone agrees on, and follow up by arguing endlessly over the numbering of exhibits and the timing of the exchange of documents. Or, fight for an exchange on the opening day of arbitration so that you can warm up for the actual arbitration by spending hours handing documents to the opposing counsel instead of putting on the case. Remember: dispute resolution is a two word term, and the first is dispute.

5. Briefs—who needs briefs?

Why should you make extra work by preparing an arbitration brief to educate the arbitrator about the facts and law? Wouldn't it just be more fun and interesting for everyone to hear about the case for the first time during the arbitration? Hey, all arbitrators are quick studies and they all welcome a challenge.

6. Reschedule the hearing a million times.

Nothing endears you to an arbitrator more than rescheduling a hearing countless times. Plus, this tactic keeps the case at the forefront of the arbitrator's attention, and keeps you in the spotlight, as the arbitrator struggles to manage his or her caseload to accommodate last minute schedule changes in your case. You don't want to be just another well-managed case, do you?

7. Show up late, don't bring your witnesses and leave your cellphone on.

Who's on time for anything anymore anyway? Fashionably late is always a better plan, and it makes you the most important person in the room when you walk in. Want bonus points? Don't tell your client where to go and when to be there. That way, you can interrupt the proceedings answering frantic cell phone calls and create an atmosphere of excited expectation about the hearing schedule. A few laughs are always guaranteed by witnesses and clients who show up without documents and present their parking ticket to the arbitrator to ask for a validation.

8. Present a spontaneous and totally unrehearsed case.

Everybody knows that arbitrators are just people, and they like to be entertained. Don't fall into the trap of putting on a tightly scripted, well planned case that focuses on the issues—it's the surest way to put the arbitrator into a deep sleep. Instead, work carefully to craft a presentation filled with amusing gaffes, time-consuming scrambles for documents, wandering lines of questioning (a sure-fire tactic for keeping the arbitrator attentive while the arbitrator tries to figure out where in the world the case is going) and, above all, repetition, repetition, repetition. There's just no better way to make your point than to repeat it, repeat it, and repeat it. And repeat it.

9. Make faces—the Three Stooges, Jerry Lewis and Jim Carrey made millions using this simple technique.

Who says that you have to sit there with a straight face while some witness on the opposing side lies his tuchas off. Pulling faces is an excellent non-verbal communication technique that effectively demonstrates to the arbitrator your skepticism about the witness's credibility. Practice your entire repertoire in advance—frowns, shakes of the head, sighs, and groans. Be creative, it will enhance your case and let the arbitrator know that you're taking a professional approach. If this doesn't work, sulk and whine a lot.

10. Never complain, never explain.

It worked for Henry Ford II, why not you? Why take the time and trouble to elucidate complex factual and legal issues? Doesn't the arbitrator get the big bucks to figure it out for himself? And don't get thrown by that old arbitrator's trick of asking you questions directly. Anyone knows that arbitrators don't really want to know the facts, they are only asking questions to keep both sides guessing about what they are thinking. Your best choice is to pretend that you didn't hear the question, or respond by asking when it's time to take a break so that you can hit the Starbucks downstairs. And closing briefs after the testimony is over? Forget about it. Once everyone has testified, it's time to move on and wait for the decision. After all, you and your lawyer have done an out-standing job—a closing brief would just be overkill. Time to sit back and wait for the award. You deserve it.

A Serious Look at Arbitration

This article was written for and published by the Association of Business Trial Lawyers of Los Angeles, CA. (Robert S. Mann, Esq., author. Association of Business Trial Lawyers ABTL Report, Volume XXVI No. 1 Fall 2003, p. 3). Reprinted with permission.

What We Learn In Arbitrator School

The first thing that one learns in "arbitrator school" is that cases are to be decided under broad equitable principles so that the result is fair and just. Although it sounds simplistic and obvious, the second thing that one learns in arbitrator school is that the arbitrator has to make a decision, regardless of the quality or quantity of proof that the parties provide in the arbitration.

After "graduation," and in the process of arbitrating cases, an arbitrator quickly learns the relationship between the first lesson and the second lesson. In some cases, the quality and quantity of proof makes it relatively easy to reach a decision, leaving the arbitrator with a strong positive feeling that a just and fair result has been determined. In others, it is enormously difficult, leaving the arbitrator with a queasy uncertainty about the appropriateness of the result.

How can knowing more about how arbitrators think make you a better advocate in arbitrations? Simply stated, a better understanding of the standards under which arbitrators decide disputes and the manner in which arbitrators approach the task of reaching a decision can prove to be invaluable in persuading an arbitrator and more effectively representing parties at arbitrations.

Let us start with some basics. In seeking to reach a resolution, an arbitrator wants to determine two things: What is the essence of the dispute? What is a fair and just result? What does this mean to the litigants and their lawyers? First, although many business cases appear to be complex, involving lengthy factual scenarios and numerous documents, the essential dispute between the parties is almost always relatively simple. Second, no matter how complex or simple the dispute, the arbitrator wants to reach a just and fair result.

How does this operate in the "real world" of arbitrations? An example will help to show how both important considerations manifest themselves in the arbitration process. Suppose the arbitration involves a dispute between the buyer and seller of a home. The buyer had deposited $75,000 into escrow. During escrow the buyer had the right to inspect and either approve or disapprove various physical and other conditions. The buyer disapproved a particular physical inspection contingency and sought to cancel the escrow and obtain the return of his deposit. The seller refused to return the deposit.

Like most real estate transactions, there would be many documents in the case, including the purchase and sale agreement, the escrow instructions, inspection reports, letters, demands, records of telephone calls and the like. There were also many arguments made by both parties on the issue.

But at the end of the day, the two analytical guideposts still direct the arbitrator toward his or her goal: what is the essential issue of the dispute, and what is a fair and just result? In our hypothetical example, the essential issue was whether the buyer's disapproval was proper and timely. What would proper mean in this context? Was the disapproval in bad faith? Did the buyer give the seller the opportunity to cure? Was there some sort of detrimental reliance on the part of the seller? The resolution of this specific issue might, and usually does depend on only one or two documents and one or two items of testimony. Likewise, and as discussed more fully below, what is fair and just may depend on a variety of factors, often not limited to the documents or the holdings of the case law.

With regard to the essential dispute, all too often, litigants and their counsel get caught up in the minutiae of the dispute, the lengthy factual background, the personality differences, the suspected motives, the hostilities, and the repeated (and sometimes self-serving) letters of counsel. The focus on these largely extraneous issues can detract from an effective and persuasive presentation for two reasons. First, it distracts the arbitrator from the task of searching for the essential dispute between the parties. Second, and particularly when one party presents more "irrelevancies" than the other, it has the effect of making the arbitrator believe that the party presenting a great volume of seemingly irrelevant material is attempting to mask the weaknesses in that party's case.

With regard to reaching a fair and just result, the issue becomes more complex, because what the arbitrator believes to be fair and just depends on many factors, some of which may have little to do with the evidence and arguments. In order to understand this, consider again the fact that an arbitrator must reach a decision, regardless of whether the quality and quantity of proof makes it easy or difficult to reach that decision. And again keep in mind that every good arbitrator wants to make the "right" decision, meaning the most just and most fair decision possible under the circumstances.

Reaching the decision, and reaching the "right" decision, depends on a host of tangible and intangible elements. First, recall that what the law requires does not always correlate precisely with a fair and equitable result. This basic fact is overlooked far too often, perhaps because lawyers are so used to dealing with highly technical issues. Thus, reliance on semantics, technicalities, hair-splitting, and Jesuitical (or, to be ecumenical, Talmudic) interpretations of the law are rarely helpful to the arbitrator's decision making process. Instead, invoking fairness and justice is likely to be more effective.

Second, while it is certainly important to direct and focus the proof toward a result of fairness and justice, the quality and the quantity of that well-directed proof is equally important. Remember that the arbitrator is evaluating two kinds of "proof" in an arbitration, although only one, strictly speaking, is actually "proof." The two kinds of "proof" are evidence and argument.

Effective evidence in an arbitration is evidence that speaks to the essence of the dispute and is directed to substantial fairness. An example will illustrate this point. Assume that the parties are arbitrating a dispute over change orders on a construction project. The contract language contains a provision that the contractor is not entitled to additional money for change orders unless the owner has approved those change orders in writing before the work was done. The contractor is claiming that additional work was done and is demanding that the owner pay for that work.

The owner might be tempted to focus the testimony on the contract, and to point out the contract terms that require written change orders. The owner might argue that a strict interpretation of the contract absolves the owner of any liability for the changes because there was no written change order. But the arbitrator wants to know more about what is fair under the circumstances than what is only set forth in the contract. The arbitrator would probably want to know, under the facts of this case, whether the contractor was trying to take advantage of the owner, or whether the owner was trying to take advantage of the contractor, considerations that are consistently important in disputes over change orders. Thus, if one represented the contractor in this dispute, one might wish to direct his or her testimony to why it was equitable to award the change order charges to the contractor despite the contract language, perhaps by pointing out that the owners were in a tremendous hurry (too much of a hurry to comply with the written change order procedure), or that the owner's agent instructed the contractor to make the changes, or that the owners were physically present and saw the changes being made (and did not object).

Effective argument is equally important, because of the arbitrator's desire to reach the right result. Effective argument does not necessarily mean a convincing verbal style (although it can't hurt to have one), nor is it limited to oral arguments made during or at the end of an arbitration hearing. Effective argument really means credibility. Establishing credibility starts from the first moment of the arbitration process. It starts with a clear and concise statement of facts and arguments in support of the arbitration claim or the response to that claim. It continues with professionalism throughout the pre-arbitration hearings, including an approach to the process that reflects knowledge of the rules. It is assisted by briefs that are dependable and carefully reasoned and that do not overstate or omit the facts, legal authorities or the arguments based on those theories.

Sometimes it only takes a seemingly small item to adversely impact upon credibility, but that adverse impact may have a lasting and profound effect on the arbitrator. A case citation that upon further examination and reading of the case leads the arbitrator to believe that the author has not correctly summarized the court's holding. A statement of facts in a brief that is not supported by the exhibits to that very same brief. A representation that a certain witness will testify to a certain thing, followed by a testimony that is far different.

All of these tangible and intangible elements combine to lead an arbitrator to feel that he or she can trust a party and counsel, or that he or she should be suspicious of that party and counsel. The arbitrator's feelings of trust are critically important to the outcome of the proceedings. Remember that the arbitrator must make a decision, no matter what the quality and quantity of proof. In a close case, trust can often make the difference.

What arbitrators learn in arbitrator school is the obverse of the trial lawyer's coin: what the lawyer and the parties need to prove and how they need to persuade. The arbitrator looks to the essence of the dispute. The successful party will tailor the evidence to the essence of the dispute. The arbitrator looks to what is fair and equitable. The successful party will direct the evidence and argument to the very same issues. Counsel who know and understand the arbitrator's concerns are substantially more likely to prevail and effectively represent parties at arbitrations.

THE LITIGATION PROCESS

What happens if your case isn't settled in mediation or doesn't go to arbitration? You will find yourself in court, engaged in a lengthy, difficult, expensive, and uncertain process. What follows is an overview of the process. Understanding what happens in a construction lawsuit is, in many ways, helpful to understanding how to avoid the problems that might result in a lawsuit.

A Typical Residential Construction Case Resolved by Mediation

Most construction lawsuits involve a combination of claims, arising from disputes over payment to allegations of construction defects. Many times a contractor's claim for unpaid money under a contract is met with a claim for construction defects.

For the purposes of illustration, let's assume that Contractor Dan, Inc., has just completed the construction of a new home for Mr. and Mrs. Potter. The contract between Dan and the Potters provides that the Potters are entitled to hold back 10 percent of the contract price until the architect signs off that everything has been built properly. In addition, the last draw request from Dan to the Potters hasn't been paid. The Potters and the architect are pleased with the appearance of the house, and everything looks good as the architect is making his inspection, but while the architect is in the process of inspecting the house, it starts to rain, and water begins to pour in through a light fixture in the living room ceiling.

The Potters panic. Their brand new dream house leaks like a sieve. The architect recommends that they call a friend of his who specializes in construction defects. The friend inspects the house and finds multiple construction defects. Dan is present during the inspection and assures the Potters that he will fix everything and that all the problems are minor. The Potters are starting to lose confidence in Dan, because their expert is advising them that he found major problems in the house.

Dan is worried because the last payment and retention represent nearly all of his profit in the job, and he still has a few subcontractors who haven't been paid. At a meeting with the Potters, their expert, and the architect, Dan loses his temper and accuses the Potters of trying to stiff him on his contractor's fee. The Potters throw Dan out of the house and tell him that he will be hearing from their lawyer.

The next day the Potters sit down with a construction lawyer and present their tale of woe. The lawyer recommends that they first try to mediate with Dan, but the Potters tell the lawyer that they don't want to be in the same room with Dan and want to go to court. After explaining why a lawsuit will cost a great deal of money and time, and why mediation would be a better approach, the lawyer is instructed by the Potters to file a lawsuit against Dan. The lawyer recommends that if they are going to court, the Potters should also make a complaint to the state agency that regulates contractors and file a claim against Dan's surety bond.

The lawyer for the Potters gets right to work and files the lawsuit that week. Dan is served with the lawsuit and consults his own lawyer. Dan explains that he is owed about

$50,000 on the last draw and about $65,000 on the 10 percent retention. Dan tells his lawyer that of the $50,000, about $30,000 is owing to several of the subcontractors, who have threatened to record mechanic's liens on the Potters' home.

Dan's lawyer explains to him that even though Dan might believe that there is nothing substantially wrong with the house, if the Potters convince a judge or jury, Dan will have to look to the subcontractor who did the work for contribution (called *indemnity*). For this reason, Dan's lawyer recommends that Dan bring the major subcontractors into the lawsuit. Dan is worried, because he knows some of these subcontractors well, and he does not want to drag them into a lawsuit, but his lawyer insists that this is the correct procedure.

Dan's lawyer prepares an answer to the Potters' complaint, a cross-complaint against the Potters, and cross-complaints against the subcontractors. The papers are filed and served on the Potters and the subcontractors. What started out as a fight between Dan and the Potters has now evolved into a lawsuit that includes 12 other parties, each with his or her own lawyers.

After receiving the papers from Dan's lawyer, the Potters' lawyer, who has substantial experience in construction defect litigation, calls Dan's lawyer to talk about the case. The two lawyers discuss the fact that with all of the subcontractors involved, the case will become very expensive very fast. The two lawyers agree to have the court issue a Case Management Order that will help everyone manage the case in a more economical way. The two lawyers also discuss the fact that although Dan and the Potters are upset and don't want to mediate the case, the lawyers both know that the court will send the case to a mediation eventually anyway, and that it would be in everyone's best interest to direct the dispute toward a mediation.

Dan's lawyer starts to prepare the Case Management Order. He wants to include the following major items:

1. The appointment of a mediator;
2. A schedule that will require the Potters to inform everyone of their claims, including each claimed construction defect, the method of repairing the defect, and the cost of repairing the defect;
3. A protocol for testing the Potters home to investigate the nature and extent of the claimed defects;
4. Strict limits on "discovery," meaning that none of the parties will be able to take depositions, send written questions called *interrogatories*, or request documents until after the mediation if the mediation does not settle the case;
5. A protocol for the exchange of information about insurance; and
6. An arrangement for everyone to deposit their documents in a central location so that everyone can review all the documents.

After the lawyers agree on the form of the Case Management Order, they present it to the judge, who signs it. The judge is pleased that the lawyers have agreed on a Case Management Order and have agreed on a mediator, because the judge has many cases on her calendar and knows that the mediator will likely settle the case, reducing the strain on the court's resources.

Dan's lawyer now serves the subcontractors with his cross-complaint and he contacts Dan's insurance company to request that the insurance company defend Dan against the complaint filed by the Potters. He does this by sending a letter "tendering" the defense of the claim to the insurance carrier.

After a month, the subcontractors begin to file their answers to Dan's cross-complaint. Many of the subcontractors contact Dan directly to complain about being dragged into the lawsuit. Dan does his best to explain why they are involved, but it puts a strain on the relationships that Dan has had with many of his subcontractors. The lawyers for the various subcontractors also contact the insurance companies and tender the defense of Dan's claims.

The Case Management Order requires Dan and the subcontractors to deposit information from their files and to answer questions about their insurance. They comply and the attorney for the Potters goes to the document depository to review the information. The lawyer for the Potters has now hired some additional experts and the experts are recommending that tests be conducted at the Potters' home. The Case Management Order has a protocol for the testing, and the attorney for the Potters gives everyone notice that the Potters' experts will be conducting tests.

Some of the testing will involve the removal of finished surfaces, such as drywall and stucco. This is called *destructive testing*. Dan's lawyer tells the attorney for the Potters that he might want to conduct some additional testing at the same time, and an agreement is reached to have enough people there to do the destructive testing and make repairs to the house immediately afterward.

The day of the testing arrives. The Potters, their lawyer, and their experts are there, along with Dan, his lawyer, his experts, and the subcontractors, their lawyers, and their experts. It's a regular circus. The Potters' lawyer has also arranged to have a professional videographer videotape the testing, because he believes that he will be able to show water cascading into the house and it will have a dramatic impact if ever shown to a jury.

All of the experts have done destructive testing hundreds of times and they know the routine. They are all armed with cameras, clipboards to make notes, and various measuring devices. The testing starts at the windows (remember that windows and doors are a prime source of leaks), and the testing crew sets up a "spray rig." Because windows are tested at the factory under controlled conditions mandated by the American Society of Testing and Materials (ASTM) and the Architectural Aluminum Manufacturers Association (AAMA), the ASTM has an established testing protocol to re-create conditions that the windows should be able to experience in the field while remaining watertight. Water is sprayed onto the windows while a few of the experts wait inside to see if water intrudes. After a few minutes, water starts to seep through onto the interior window sills and the spray rig is turned off. Now it's time to figure out why the windows are leaking.

Figure 10-1 shows some real destructive testing of a door assembly, using blue dye to show the path of the water intrusion.

Dan has been watching the process and his heart starts to sink as he sees the water pond up on the window sill. Dan shoots his window installer a dirty look and gets a glum look in return. The Potters' experts carefully break away the stucco around the window to discover whether the problem lies in the installation or the window itself. After breaking

Figure 10-1

Destructive testing of a door assembly

the stucco apart, they carefully peel back the builder paper to reveal the flashing material. As soon as the flashing material is exposed, the problem becomes obvious: the flashing was not installed correctly. It was reversed lapped, meaning that the flashing across the head of the window wasn't put on first; it was put on last, and water is running into the gap between the horizontal piece of head flashing and the vertical flashing on each side of the window.

Dan's window installer subcontractor tells everyone that it must be a single mistake. The Potters' expert suggests that they test this theory and open up another window. They do so and find the same condition. It appears that the laborer assigned to the job did not understand the proper application procedure and wasn't well supervised. Further testing reveals many other problems with the house, some technical and easily correctable and some as serious as the window problem. Dan feels like he is going to sink into the dirt and he's furious at his window installer. Mr. Potter looks as if his mood is swinging between dejected and murderous. The subcontractors look like they need a place to hide.

By now it's the end of the day and the workers are making efforts to patch up the areas that were tested. Dan takes his lawyer aside for a conference and tells him that he doesn't know what to do. Dan feels like he has already lost the case. Looking over his lawyer's shoulder, Dan sees the Potters standing next to each other watching the workers patch the stucco around the windows. The fact that they are upset is obvious to everyone.

Dan decides to take matters into his own hands. He tells his lawyer to wait and he walks over to the Potters and asks them to speak to him alone for a moment. Dan tells the Potters that he feels terrible and he's sorry. He can't understand how these problems happened. He wants to stand by his work and he sees that it's obvious that substantial repairs will be necessary. Dan tells the Potters that he wants the subcontractors to step up and make the repairs or pay money so that someone else can fix the problems. Dan says that his lawyer has told him that mediation is the best way to get these kinds of disputes resolved and that maybe everybody should think about going there right away and not spending more money on the lawyers.

The Potters are angry, but they can see that Dan is making a sincere effort to do the right thing and take responsibility for what happened (never underestimate the power of the words "I'm sorry"). They promise to talk to their lawyer and have him call Dan's lawyer right away. They thank Dan for being up front and not trying to avoid the blame. When everyone walks away, they feel a little better about the future and a little more hopeful that the situation won't be a total disaster. Dan's "mea culpa" was a good move.

The next day the Potters' lawyer calls Dan's attorney and tells him that the Potters really appreciated Dan's forthright reaction. He suggests that he get working on a cost of repair and that they call the mediator and get an early date for a mediation. Dan's lawyer agrees and calls Dan to give him the news.

The Potters' lawyer and their experts start work on the repair numbers. In order for the case to be mediated effectively, the Potters need to prepare a document for Dan and the subcontractors that contains a *scope of repair*, meaning an identification of where all the defects are located and where repairs will be made; a *method of repair*, meaning a description of how the repair is to be made; and a *cost of repair*. This is usually called a preliminary defect list and cost of repair estimate, and it must be sent to and reviewed by Dan, the subcontractors, the lawyers, and the experts well before the mediation so that everyone can be prepared to discuss the claims at the mediation.

When the document is published to Dan and the subcontractors, the cost of repair is more than $350,000, even though the original job cost only $650,000 and not everything in the house is defective. Although Dan is dismayed when he sees this figure, his lawyer and the experts aren't surprised. They know two important things about the Potters' number: that it is much more expensive to fix a structure than build it in the first place, and that the Potters' experts will try to include every conceivable claim and use the highest possible costs of repair to extract the greatest amount of money from the settlement. They also know that during the mediation process the numbers will be tested and that the ultimate settlement probably will be less than the estimated cost of repair.

While the Potters have been working on the defect list and cost of repair, Dan's lawyer and the lawyers for the subcontractors have been fighting their own battles with the insurance carriers. Dan has a Commercial General Liability (CGL) Policy, with all the appropriate policy provisions, but his insurance company is taking the position that it will not defend Dan in the lawsuit because there is no "resultant" damage to the Potters' home. The insurance company argues that because the defects were caught before they did any real damage, there is no resultant damage and the company won't pay for "mere construction defects." Dan's lawyer recommends that they hire an insurance coverage lawyer, an attorney who specializes in disputes between insurers and insurance companies regarding coverage issues.

The coverage lawyer reviews the file and writes the insurance company a stiff letter. In the letter he reminds the insurance company that their obligation to defend Dan is much greater than their obligation to pay a judgment if the Potters obtain a judgment. He also threatens to sue the insurance carrier for "bad faith" for refusing to defend Dan. About a week later, the insurance carrier writes back and agrees to defend Dan, but only with a long and detailed reservation of rights.

Dan asks his coverage lawyer to explain the reservation of rights and he is told that the insurance company is reserving the following important rights:

1. The right to withdraw from defending Dan if at any time they conclude that the Potters' claims in no way could conceivably be covered;
2. The right not to pay any judgment that the Potters might obtain; and
3. The right to recover the attorneys fees and costs that the insurance carrier will pay during the defense of the case.

Dan is pretty worried by this explanation, but his lawyer tells him that as a practical matter, the case is likely to be settled with a contribution from the insurance carrier, a contribution from the subcontractors' insurance carriers, a contribution from Dan, and one from the subcontractors. When the case is settled, the insurance company will close its file and the case will be over.

During the next couple of months, the defense side of the case reviews the Potters' defect list and cost of repair estimate and develops their own response. The subcontractors work through their own insurance issues. Finally, the parties obtain a mediation date from the mediator.

A week or so before the mediation, the lawyers submit written statements, called *mediation briefs*, to the mediator. The mediator has considerable experience in construction disputes, so the lawyers do not have to re-invent the wheel, but they take care to explain the various claims and the reasons why it won't cost as much as the Potters claim to fix the house. They also make the mediator aware of the insurance problems and that there is a negative "wrinkle" in the case: one of the main subcontractors has not been able to convince its insurance carrier to defend.

On the day of the mediation, everyone meets at the mediator's office. The Potters, their lawyer, and their lead experts are there. Dan, his lawyer, his lead expert, and his insurance carrier representative are there. The subcontractors are present with their lawyers, experts, and insurance claims representatives. The mediator, Michael Mitchell, having read everyone's briefs, has a good handle on the case and is ready to proceed. The mediator has everyone sign a mediation confidentiality stipulation before they start. Under the terms of this stipulation, everything that is said in the mediation remains strictly confidential and cannot be used in court if the parties fail to reach a settlement. The mediator also explains that anything that is said to him in a private session with one party, without everyone else being present, will remain confidential. This is done so that the parties can have a frank and open discussion with one another and with the mediator.

Mediator Mitchell likes to start construction mediations with a "Plaintiff's Dog and Pony Show." This is a procedure where the Potters' experts conduct a presentation, usually using PowerPoint or slides of the defects, to illustrate the probable cause of the defects, the scope of the defects, the proposed method of repair, the proposed scope of repair, and the estimated cost of repair. This is helpful to define all the issues and focus everyone on the important points.

After this, Mediator Mitchell meets separately with Dan. The subcontractors and the Potters are in other rooms, waiting their turns to talk to the mediator in private. The mediator wants to hear Dan's take on the Potters' presentation. Dan and his experts admit that most of the claimed defects actually exist. He protests that some of the claims are inconsequential and have been included merely to force the numbers up. Dan's main problem is the cost of repair. Dan feels that the Potters have been persuaded by their lawyer to let their expert run amok on the cost of repair. Dan thinks that every single problem could be repaired for less than $50,000. Dan also reminds the mediator that he is owed a substantial amount of money on the last draw and the retention.

The mediator, Dan, and Dan's lawyer strategize for a few minutes about what contributions they want from the subcontractors. The mediator makes a list of the subcontractors with a few columns, including his "Demand" column, his "Probable" column and his "Offer" column. This helps the mediator keep track of the conversations with the subcontractors.

The mediator then checks in with the Potters and tells them that he will be spending a good deal of time with the subcontractors. He explains that the bulk of the money to settle the case will come from the subcontractors. He also asks the Potters to think about Dan's claims on the last draw and the retention and reminds them that some resolution of those issues will have to be achieved if there is to be a settlement.

The mediator then starts a lengthy process of discussing the claims with each individual subcontractor in separate meetings, called *private caucuses*. By late afternoon, the mediator has a better idea of who is going to contribute what. He then meets with Dan and his team to go over the subcontractor numbers, reminding everyone that Dan and his insurance carrier will have to make up the difference between the ultimate settlement figure and the contribution from the subcontractors.

The numbers are still pretty far apart. The mediator has raised about $95,000 from the subcontractors, but he knows that he still has a long way to go. Some of the subcontractors aren't well prepared for the mediation, and they need more time to go back and discuss the claims with their insurance companies.

The mediator then meets with Dan's team and the Potters and their team. He explains that he has made progress, but they have more work to do. He also explains that it is fairly typical in a case like this to have the first session be devoted more to education than negotiation, even though everyone tries hard to educate all the subcontractors before the mediation. The mediator suggests that they pick another date in about 30 days to try to wrap up the settlement.

In the intervening month, Dan's team supplies the subcontractors with additional information, and at the mediator's suggestion, some direct conversations occur between the experts, without the lawyers. The mediator knows from experience that many times the experts can reach agreement if they don't have to perform for the lawyers and the clients, particularly because many of the experts see one another in case after case and they have a good working relationship.

At the second mediation session, everyone is more of a mind to get the case settled. Dan's insurance carrier arrives willing to make a substantial contribution toward a settlement. The subcontractors are acting more reasonably. By early afternoon, the mediator has a bundle of money from Dan's group and the subcontractor group of $225,000. But Dan wants nearly $100,000 on his last draw and retention.

The mediator takes this collective offer into the Potters. The Potters and their lawyer are disappointed. They want to see a number starting with a *3*. The mediator reminds the Potters that they still owe Dan a considerable sum for the last draw and retention. The mediator spends some time discussing with the Potters the realities of the litigation process, namely that it is very time-consuming, uncertain, and expensive. The mediator points out that because there are so many parties in the lawsuit, the cost of formal discovery alone could exceed $100,000. He reminds the Potters that they might not win on every claim, and that if the jury thinks that they are being greedy, they might not win at all. The mediator suggests that the real problem in settling the case is that Dan's insurance carrier will be putting up money to pay the defect claims, but the money to pay Dan's claims must come directly from the Potters.

The mediator suggests that the Potters consider a settlement where he tries to raise $250,000 and Dan reduces his claim to $50,000. This would "net" the Potters $200,000, less their attorneys fees and costs. The Potters are worried about the fact that their experts have stated that it will cost $350,000 to fix the house. The mediator casts a meaningful glance at the Potters' expert. Recognizing the cue, he explains to the Potters that there are other, less expensive ways to fix the problems, and that everything could probably be repaired for substantially less. After some discussion, the Potters agree to try for the $250,000 settlement.

By the end of the day, the mediator has raised the $250,000 and has obtained Dan's agreement to take $50,000 on his draw and retention. The mediator writes up a memorandum of settlement that is signed by everyone in anticipation of the lawyers preparing a formal settlement agreement, and everyone signs. Dan and the Potters exchange handshakes and Mrs. Potter ventures to give Dan a little hug. She thanks Dan for being a gentleman and everyone expresses regrets that the whole thing happened.

Dan and the Potters leave the mediation and each one celebrates that night, knowing that the dispute is over and they can go on to more productive things.

Within 30 days, the money is paid and the lawsuit is dismissed. The Potters start the process of repairing their home.

When No Settlement Occurs

Let's replay our scenario, with one important difference: the mediator cannot bridge the gap between the Potters' demands and the defense offers. The case doesn't settle. Now what happens?

The parties were assigned a date by the court to return to court and tell the judge whether the case settled at the mediation. They dutifully show up in court and report that the mediation was not successful and they doubt whether additional mediation sessions would be helpful. The parties request that the court "open up" discovery and that the court set a trial date. The court reluctantly does both but sets the matter for a Mandatory Settlement Conference (a process much like mediation, but conducted by a judge, not a mediator) a few weeks before the trial.

Now the parties start the real litigation process, which is largely one of formal discovery. Dan's lawyer informs everyone that he intends to take the depositions of all the PMKs (the Persons Most Knowledgeable) of the subcontractors, along with all the subcontractor experts. The Potters want to take Dan's deposition, along with his experts. A discovery free-for-all starts, and it begins to consume tens of thousands of dollars of attorney fees and costs (costs include court reporter fees, expert fees, and the charges for deposition transcripts).

The parties and their lawyers spend scores of days in depositions. All of the PMKs testify. The Potters and Dan testify, along with their experts. The parties exchange documents, including reports of experts. The defense side asks the court for permission to conduct more destructive testing, and the court allows this. In the course of tearing up the Potters' home to conduct this additional destructive testing, more defects come to light, which is unfortunate for Dan and his subcontractors.

As the trial date approaches, the parties prepare to attend the Mandatory Settlement Conference (MSC). Everyone appears on the appointed day and they are assign to a judge who specializes in helping litigants reach settlements. The lawyers meet with the judge first, and she wants to know why the case did not settle at the mediation. The lawyer for the Potters tells the judge that Dan and his subcontractors aren't being realistic about the case, that the case was good at the time of the mediation but that it has just gotten better since, with the additional destructive testing. The defense team tells the judge that the Potters are exaggerating. They point out that they could repair all the claimed defects for less than half of what the Potters experts say it will take.

The judge meets with the Potters' counsel alone. A few minutes into the conversation, it is obvious to the judge that there is a major problem in settling the case: the cost of litigating the case has been huge, and if the case settles, the Potters probably won't obtain enough money to make the repairs. Since the time of the unsuccessful mediation, the Potters' side alone has spent more than $150,000 in attorneys fees and another $50,000 in costs, including the costs of their experts. If the case settled for the same $250,000 that was under discussion at the mediation, the exercise would be pointless. The judge sends away the Potters' lawyer and spends some time with the defense.

Dan's lawyer and Dan's insurance carrier claims representative meet with the judge. The judge talks to them about the exposure in the case, meaning the amount that Dan would have to pay if he lost the case. The carrier claims representative tells the judge that the Potters want so much money that Dan might as well try the case: the Potters are not giving Dan and his insurance company much of an incentive to settle. The claims representative points out that everyone has already spent most of the money anyway, and there's not much more that everyone will spend between the MSC and the end of the trial, even if the trial takes a few weeks.

The judge spends most of the day trying to settle the case, but it's obvious to her that the best time to have settled the case would have been at the mediation, and the parties lost their best opportunity to make the numbers work. Now, too much money has been spent on a case where the facts haven't changed all that much. The settlement cannot occur unless the Potters drastically reduce their demand. Even after a long discussion with the Potters, the judge cannot convince them to do so. At 4:00 p.m, the judge announces to everyone that she cannot settle the case and advises everyone to prepare for trial.

In a flurry of last-minute activity, the parties get ready to try the case. Because construction defect cases are almost entirely decided on expert testimony, the lawyers and the experts spend a huge amount of time getting their presentations ready. In the background, the lawyers prepare briefs to the trial judge, proposed jury instructions, and prepare motions to be heard at the outset of the trial to try to limit claims and defenses.

The day of the trial arrives. The Potters, Dan and the subcontractors, and their lawyers are all present. The first day is spent with the trial judge hearing the pre-trial motions. The judge orders a jury panel for the next morning. The next morning and for 21 more mornings after that, the parties are in trial. The Potters and Dan are there every day, because their lawyers have told them that unless they are personally present the jury is likely to conclude that they don't care about the outcome of the case. As the days go by, they become increasingly exhausted. Both the Potters are losing time away from their jobs. Dan is in the middle of another job and has to make arrangements to have someone else supervise the work while he's gone.

Finally the trial drags to a close. The final arguments are made. The Potters ask the jury for $450,000. Dan's lawyer tells the jury that everything could be fixed for $125,000 and that Dan should be awarded $95,000 on his contract. After two days of deliberations, the jury comes back with an award in favor of the Potters for $260,000 and they award Dan $45,000 on his claims. It's almost the same amount that everyone was discussing at the time of the mediation. The jury found mostly in favor of Dan on his cross-complaint against the subcontractors, awarding him $230,000 and "deducting" $30,000 for Dan's failure to supervise the work adequately.

After the trial, the parties and their lawyers interview the jurors in the hallway. The Potters' lawyer wants to know why the jurors awarded only $260,000 and not the whole $450,000. The jurors tell the Potters' lawyer that they thought the experts were just "hired guns," and that it was obvious to them that the experts had been instructed to make the numbers as high as possible. By the same token, they also didn't believe Dan's experts, because they felt that Dan's experts had been hired to bring in low numbers. They simply tried to get to a number somewhere near the middle because they thought that there were problems at the house that needed to be fixed.

The lawyers return to court a few weeks later to argue the issue of attorneys fees and costs. The fees were $200,000 and the costs $75,000 through the trial. Because the Potters won the case, they were considered to be the "prevailing party" for the purposes of awarding attorneys fees under the contract. The trial judge had the discretion to reduce the attorney's fees because the contract provided for the recover of "reasonable" attorneys fees. Dan's lawyers argue that the lawyers for the Potters had made the case unnecessarily expensive. The judge ended up reducing the attorneys fees to $145,000 and awarded all the costs. The total judgment that was entered by the court in favor of the Potters and against Dan and the subcontractors was $480,000.

The Potters owed their lawyers $275,000 in fees and costs, leaving them with $205,000 to fix their home.

Dan and his lawyers send the judgment to Dan's insurance carrier. A week later, the carrier wrote back. The carrier took the position that none of the claims in the judgment were covered by the policy and the carrier, having fulfilled its duty to defend Dan, now declined to pay any portion of the judgment. Dan and his lawyer were told by the subcontractors that their carriers were taking a similar position. Of the subcontractors against whom Dan obtained judgments on his cross-complaint, only one or two were financially able to respond to the judgment. The rest were "mom and pop" type operations, who lived from job to job, and had no assets from which a judgment could be satisfied.

It was obvious to Dan that he would be left with the responsibility of satisfying a judgment for more than $400,000. Dan has followed the advice of his lawyer over the years and has carefully kept his construction company separate from his personal business. The construction company is a corporation and the corporation has no assets of its own except for Dan's pickup truck. Dan and his lawyer discuss the option of filing for bankruptcy for the corporation. They agree that they will discuss with the Potters the idea of a post-judgment settlement, and if they can't work out terms, that Dan will bankrupt the corporation.

The Potters are not happy with the fact that they will "net" only about $200,000 from the case, but they are wild when they find out that Dan could bankrupt his company and they might get nothing. They tell their lawyer to make the best deal possible. Because a bankruptcy will cost Dan a considerable amount in legal fees and will make it difficult for him to start up again in business, and because Dan can get some significant money from the subcontractors, Dan works hard to try to put a settlement together. After some weeks of negotiation, a total package of $135,000 is presented to the Potters with a "take it or leave it' provision.

The Potters are hugely disappointed but feel that they have no alternative but to take the settlement. The Potters end up with no money left to fix their home, because all of the settlement funds went to pay legal fees and costs. Dan ends up with nothing for the money that he was owed.

The trial has been a bad alternative for all parties. A month or so later, Dan runs into Mr. Potter at the home center. Almost at the same time, they both tell one another: "We should have settled at the mediation." Had they done so, the Potters would have ended up with enough money to fix the house because Dan's insurance carrier would have contributed to a settlement, Dan would have received something on what he was owed, and the matter would have ended up productively for both parties. The time and expense of the litigation process consumed the money available, and everyone lost.

_11_ Conclusion

When I started to write this book, I intended it to be a simple list of common construction defects with some advice on how best to avoid them.

As the book developed, it occurred to me that both owners and contractors would be better able to avoid not only defects but also conflicts if they more deeply understood the process of construction and were more aware of what usually goes wrong. It also occurred to me that owners and contractors could use some guidance about what to do when things do go wrong.

In retrospect, after having described all of the usual suspects of construction litigation, including the background to each, the cause of each, and the way in which to avoid each, they seem surprisingly simple. And, in fact, I think they are simple. Each of the construction defects is just a part of the normal construction process. There's nothing particularly exotic about them. In fact, it doesn't take any enormous skill to avoid making the mistakes that cause the defects. It does require awareness and a willingness to pay attention to detail. It requires a change in attitude from "I don't know and I don't care," to "I know and I care."

I hope that this book will serve the interests of owners and contractors and enable them to build better buildings, to avoid conflict, and to have a more profitable and more enjoyable relationship during construction—one of the great adventures and challenges of life.

Paul Practitioner
1234 Main Street
Anytown

Attorneys for Plaintiff Homeowner

Superior Court of the State of California for the County of Los Angeles

	)	[PROPOSED] CASE	
	)	MANAGEMENT ORDER	
Unhappy Homeowners,	)	Case No. ____	
	)	Judge ____	
	)	Dept. L	
Plaintiffs,	)	Complaint Filed:	
	)		
v.	)	Status Conf.:	
Ajax Construction and DOES 1-100, inclusive,	)		
	)		
	)		
Defendants.	)	Trial Date:	None Set
	)		

1. **General Provisions.**

    1.1 Purpose. This construction defect action is deemed complex under California Rules of Court, Rule 1800, in that it involves a large number of parties and claims, and trial, if it occurs, is likely to be prolonged. This Case Management Order ("CMO") is entered to reduce the costs of litigation, to assist the parties in resolving their disputes if possible, and if not, to reduce the costs and diffi-culties of discovery and trial.

    1.2 Code Governs Where Silent. On any matter as to which this CMO is silent, the Code of Civil Procedure, other statutes, the California Rules of Court, and the local rules of this Court shall be controlling.

    1.3 Confidentiality of Proceedings. Evidence Code sections 1119, et seq. and 1152 apply to all mediation sessions, settlement conferences, and formal or informal expert meetings. Any final settlement agreement made during a Court ordered settlement conference or mediation session, as approved in writing or by appearance on the record by the parties to the settlement, shall be admissible in any motion for good faith settlement or action or proceeding to enforce the settlement, including, without limitation, a motion under Code of Civil Procedure sections 664.6 and 664.7. Any final settlement agreement made during a settlement conference or mediation session shall comply with Evidence Code sections 1118, 1123 and 1124.

2. **Parties**.

2.1 Naming Additional Defendants and Cross-Defendants. All parties may name additional Defendants and Cross-Defendants, who have not appeared in this action. Those parties shall be named by the date indicated in the attached CMO Time Line (attached hereto as Exhibit A), unless extended by order of the Court for good cause shown.

2.2 Certain Cross-Actions Deemed Filed and Answered. All Defendants and Cross-Defendants shall have been deemed to have filed and served cross-complaints against all other Defendants and Cross-Defendants for contribution, implied and equitable indemnity, apportionment of fault, and for declaratory relief with respect to same, except that (a) as to those parties as to whom a Certificate of Merit is required pursuant to Code of Civil Procedure sections 411.35 and 411.36, no cross-complaint is deemed filed and served until such Certificate(s) has or have been filed; and (b) as between parties represented by the same counsel, no cross-complaints shall be deemed to have been filed or served. Once deemed filed, all such cross-complaints are deemed to have been generally denied, and all applicable affirmative defenses deemed to have been raised. Any party may opt out of the deemed cross-complaint provision by giving notice of same to all parties. Any party wishing to claim breach of contract, breach of warranty or express indemnity must file and serve appropriate pleadings.

2.3 New Parties. Any party bringing in any new party shall serve a copy of this CMO upon said party with its operative pleading, whereupon this CMO shall bind such newly appearing party unless the Court grants the party relief therefrom upon noticed motion for good cause shown.

2.4 Default for Failure to Timely Respond. All parties are ordered to take the default of any newly served party if no responsive pleading is filed by the sixtieth day following service on any such new party.

3. **References**.

3.1 Settlement Referee: To facilitate settlement discussions, the Court finds that because of the complex technical issues, numerous parties and the lengthy trial if not settled, the designation of a Settlement Referee is appropriate. The Court appoints _____, Esq. as the Settlement Referee pursuant to California Code of Civil Procedure section 187 to mediate and conduct settlement conferences in this case.

1. The address and phone number of the Settlement Referee is: _____

2. The hourly cost for the Settlement Referee is his usual and customary rate.

3. No party has established an economic inability to pay a pro rata share of the fees of the Settlement Referee.

4. The fees of the Settlement Referee shall be paid one-third by the Plaintiff, one-third by the developer, and one-third in equal parts by the Cross-Defendants.

4. **Discovery**.

4.1 Stay of Discovery. All discovery, except for that discovery required or permitted under this CMO, is stayed except by Order of the Court for good cause shown.

4.2 Third Party Discovery Permitted. Discovery shall not be stayed as to depositions and Public Records Act requests of individuals and entities not a party to this litigation. A copy of any document obtained through such third party discovery shall be deposited in the document depository within thirty days of its receipt, with notice to all parties as provided below with respect to the deposit of such documents.

4.3 Document Depository. A document depository shall be established at

_____

_____

4.3.1 Documents to be Deposited. Thirty-five (35) days after a party's appearance or service of the CMO, each party shall provide to the depository the original or Bates stamped copies of all non-privileged documents, including oversized and color documents, relating to the subject real Property, as described in Exhibit B attached hereto. If any party discovers additional documents, other than documents received from the depository itself, they shall be deposited within thirty days of the party's receipt of the additional documents. The term "document" shall have the same meaning as the term "writing" defined in Evidence Code section 250, and is meant to be all-inclusive. All parties are under a continuing obligation to deposit all non-privileged documents discovered after the initial production. In the event that a party subsequently discovers documents, that party shall deposit said documents and follow the same procedure set forth in this order with respect thereto. A copy of all pleadings and discovery shall also be served upon the depository so that later-appearing parties may obtain same from the depository rather than the other parties.

4.3.2 Copying and Indexing. If originals are provided, the depository shall copy the documents, numbering each page, and index the documents. Each document's number shall be preceded by a unique two-letter code identifying the depositing party, and the code for each party is to be established by the depository. If a Bates stamped copy is provided, such party shall also provide an index.

4.3.3 Notice of Compliance. When a party's documents are copied, indexed and deposited, said party shall promptly serve on all parties a Notice of Compliance to which the party shall attach the index of documents deposited, and the date the documents were deposited.

4.4 Privileged Documents. Any party withholding any document(s) on grounds of privilege shall deposit and serve upon all parties a log listing the author, all recipients, the date, a specific description of the document(s) sufficient for a Motion to Compel, and the privilege claimed.

4.5 Inspection of Originals. For reasonable cause, any party may request an opportunity to view the originals of any document in the depository. After a reasonable attempt at informal resolution, any party may make a motion for an opportunity to view originals, for good cause shown.

4.6 Costs of Compliance. The depositing party shall bear the costs of compliance with the deposit obligation. The fees for maintaining and utilizing the Depository shall be set by the custodian of same. Fees for all parties shall be the same.

4.7 Insurance and Scope of Work Interrogatories. Thirty-five (35) days after a party's appearance or service of the CMO, each Cross-Defendant shall respond under oath to the Insurance and Scope of Work Interrogatories attached hereto as Exhibit C.

4.8 Discovery Disputes. The Court will act as Discovery Referee, and all discovery disputes shall be directed thereto. With respect to objections to any discovery, the responding party must initiate a meet and confer with the propounding party within ten days of receipt of the discovery at issue. If the objections cannot be resolved through the meet and confer process, the responding party is to schedule a telephone conference with the Court. The responding party shall give the Court and all parties at least three (3) days' notice of the conference call, which notice shall include a summary of the issues in dispute.

4.9 Discovery by Leave of Court. After notice and an opportunity to be heard, discovery additional to that permitted under this CMO may be permitted to any party, for good cause shown.

4.10 Expert Designations. Expert witness designations shall be exchanged on or before 5:00 p.m. on the date indicated in the CMO Time Line attached hereto. Any subsequent designation shall be made pursuant to Code of Civil Procedure section 2034.

4.11 Protocol for Depositions. Depositions shall proceed in the following order: Plaintiff, then developers' persons most knowledgeable and employees and representatives, subcontractors' persons most knowledgeable and employees and representatives, then all experts. Should mediation fail and the discovery stay be lifted, and party may take the deposition of the Plaintiff, other percipient witness, or expert not yet deposed. The foregoing order of deposition is without prejudice to the parties noticing the deposition of third party percipient witnesses.

4.12 Protocol for Expert Depositions. The order for expert depositions shall be by trade or discipline beginning with Plaintiffs' expert in each discipline followed by developers' expert and then subcontractors' expert. Ten (10) days prior to each experts' deposition, that expert shall deposit into the depository a complete color copy of the expert's job file and notice of such deposit shall be provided to all parties.

5. **Preliminary Defect List**.

Plaintiffs shall deposit in the depository a Preliminary Defect List by the date indicated on the attached CMO Time Line. Said Preliminary Defect List is without prejudice to being subsequently amended, need not be verified, shall have no evidentiary impact, shall be protected by Evidence Code sections 1119, et seq. and 1152, and is merely for informational purposes. This list shall contain a description of each defect alleged in the Development, including quantity and location, if known.

6. Final Defect List and Cost of Repair Estimate. Plaintiffs shall serve on all parties and deposit into the document depository a Final Defect List, including a Cost of Repair Estimate, on or before the date set forth on the attached CMO Time Line. The Final Defect List shall include a comprehensive description of all defects alleged by the Plaintiff, and an identification of the location of the defects enumerated within the Final Defect List (with sufficient particularity to allow parties to locate each defect).

The Cost of Repair Estimate shall include a comprehensive break down of Plaintiff's proposed scope of repair for each defect enumerated in Plaintiff's Final Defect List, and a comprehensive break down of costs necessary to effectuate the proposed scope of repairs. This break down shall include all general and administrative costs associated with effectuating Plaintiff's proposed scope of repair.

The Final Defect List and Cost of Repair Estimate need not be verified. Should Plaintiff thereafter amend the Final Defect List or Cost of Repair Estimate, all parties shall have an opportunity to perform additional inspections and testing related to any defects added to the Final Defect List or Cost of Repair Estimate. If the parties cannot agree on the schedule and scope of the additional inspections and/or testing, the matter shall be submitted to the Court.

The Final Defect List and Cost of Repair Estimate shall be subject to Evidence Code section 1119 and shall be used for settlement purposes only. However and in the event this matter is not resolved in its entirety by way of settlement, Plaintiff's Final Defect List and Cost of Repair Estimate shall be deemed to be Plaintiff's final statement of claims and damages for purposes of trial upon the commencement of expert depositions. From that point forward, Plaintiff's Final Defect List and Cost of Repair Estimate shall not be made inadmissible by way of Evidence Code sections 1119 or 1152.

7. Inspections and Destructive Testing by Plaintiffs. Prior to the publication of the Final Defect List and Cost of Repair Estimate, Plaintiff may conduct visual inspections and destructive testing. Plaintiff shall not conduct any further inspections or testing after the publication of their Final Defect List and Cost of Repair Estimate absent notice to all parties.

At least two (2) weeks prior to the commencement of any visual inspections and/or destructive testing, Plaintiff will provide notice to all parties setting forth the addresses and locations which will be inspected or tested, the type of inspection and testing involved, and the estimated time to complete the inspections and/or testing. Cost related to Plaintiff's destructive testing shall be borne by the Plaintiffs. The Defendants and Cross-Defendants may attend and observe Plaintiff's inspections and destructive testing, including taking field notes, photographs, and videotapes. The Defendants and Cross-Defendants may not, however, remove samples, direct openings or in any manner disrupt or delay the testing process. If they do so, they will be required to share in the costs associated with Plaintiff's testing in an amount to be agreed upon by the parties which is proportionate to the Defendants' and/or Cross-Defendants' involvement. If they are unable to agree upon an amount, the matter will be submitted to the Court.

8. Defense Visual Inspections and Non-Invasive Testing. By the date indicated in the CMO Time Line, Defendants and Cross-Defendants shall submit to counsel for the Plaintiff and serve on all parties the visual inspection/invasive testing request attached hereto as Exhibit B, listing by location those areas which they wish to

visually inspect. Plaintiff shall then publish a visual inspection schedule by the date indicated on the attached CMO Time Line, and visual inspections will occur during the time period set forth in the attached CMO Time Line. No party, counsel for any party, or expert or agent thereof shall enter into any discussions whatsoever with any resident or staff at any time during such visual inspections and non-invasive testing, or any future interior or exterior inspections.

9. Defense Destructive Testing. By the date indicated on the attached CMO Time Line, all parties shall provide counsel for the Plaintiff and serve on all parties the visual inspection/invasive testing request attached hereto as Exhibit D, listing by location the areas which they wish to perform invasive testing, including a general description of the nature of the testing. If necessary, a meeting shall take place on a date to be later determined, between counsel for the Plaintiff and the developer Defendants to determine the scope and schedule of destructive testing. Counsel for the developer Defendants shall be primarily responsible for coordinating the testing requested by the various Defendants and Cross-Defendants. Plaintiff shall then publish an invasive testing schedule by the date indicated in the attached CMO Time Line.

Costs related to defense invasive testing shall be allocated equally amongst the participating Defendants and Cross-Defendants. Participation shall be defined as the attendance and observation of the testing by the party, or its attorneys, experts, or insurance carrier representatives. The Plaintiff may attend and observe the defense destructive testing, including taking field notes, photographs, and videotapes. The Plaintiff may not, however, remove samples, direct openings or in any manner disrupt or delay the testing process. If they do, they will be required to share in the costs associated with the defense testing in an amount to be agreed upon by the parties which is proportionate to the Plaintiff's involvement. Any and all disputes regarding payment for invasive testing shall be submitted to the Court.

The party conducting the destructive testing shall promptly repair, restore and clean the Property dismantled or exposed during the invasive testing. The repairs shall be made in a workmanlike manner at the sole cost and expense of the testing and participating parties.

10. Repairs. Repairs relating to any allegedly defective condition that is the subject of this litigation may only be performed by the Plaintiff after providing at least five (5) days notice to all parties, unless emergency conditions require repairs which prevent such notice. If circumstances require an emergency, Plaintiff's counsel will provide as much notice as possible under the circumstances. In any event, Plaintiff's counsel is obligated to notify all parties upon learning of such emergency, including a detailed description of the nature of the emergency, the type of repair contemplated, the name of the contractor or other trade performing the repair and the date of the repair. Parties may inspect and document such repair upon providing notice to the Plaintiffs.

11. Settlement Conferences. Settlement Conferences shall be held on the dates set forth on the attached CMO Time Line. All activities and discussions occurring during the course of settlement conferences are protected by Evidence Code sections 1119 and 1152. Follow up settlement conferences may be scheduled at the discretion of the Mediator.

A settlement demand shall be served by Plaintiff on the developer at least forty-five (45) days prior to the date set for the first settlement conference. The developer shall serve individual settlement demands on all Cross-Defendants no later than fifteen (15) days prior to the first settlement conference.

All parties shall be required to appear with their respective attorneys, insurance carrier representatives with full policy limits authority, and any other persons with decision-making authority regarding settlement. At each scheduled settlement conference, all parties required to attend must be prepared to fully discuss settlement of all defects alleged to arise from that party's activities. Each party's representative, including insurance carrier representatives, must appear for settlement conferences in person, unless the Settlement Referee authorizes a telephonic appearance for good cause shown at least twenty-four (24) hours prior to the scheduled settlement conference. Failure to adhere to this provision may give rise to a recommendation by the Settlement Referee for sanctions against the offending party. Parties whose representatives have been authorized to appear telephonically must serve written notice to counsel for Plaintiff and the developer immediately upon obtaining authorization to do so.

12. **Miscellaneous**.

12.1 Communications with Settlement Referee: The Settlement Referee encourages that any communications which may need to be brought to his attention may be raised on an informal basis. The Settlement Referee does not wish to be on a master mailing list of the parties in this matter. However, documents or correspondence which relate to issues which are the subject of his appointment or the substitution of any counsel or dismissal of any party shall be copied to him. All moving papers relating to motions or matters to be heard by the Settlement Referee should be sent directly to him and duplicate copies need not be filed with the Court or the department to which the case is assigned.

Parties who fail to review this and other Orders of the court and the Settlement Referee and file superfluous pleadings may be subject to the imposition of sanctions. This particularly refers to cross-complaints and answers thereto deemed not required by virtue of this CMO and requests for documents or discovery relating to items deposited in the document depository.

12.2 Protective Orders. Any and all protective orders, including filing records under seal, shall comply with California Rules of Court, Rule 243.2.

12.3 Service of Documents. All documents, including law and motion, may be served via facsimile or by electronic transmission. Service by facsimile or electronic transmission shall be governed by Code of Civil Procedure section 10138). Parties shall supply all other parties with their facsimile number and e-mail addresses upon the later of their first appearance in the case or within thirty (30) days of the entry of this Order.

12.4 Inadvertently Produced Documents. Any party receiving a document which that party has a reason to believe was inadvertently produced must notify the producing party to ascertain whether the document was intended to be produced. If it was not, the receiving party must immediately return the document(s) to the sending party and treat it as if it was never produced. Upon notice from the producing party, all other recipients of the document(s) must immediately return the document(s) to the sender and treat it (them) as if never produced. In so doing, all parties are to comply with all applicable rules of professional conduct.

12.5 Relief from CMO. Any party, upon application to the Court, may seek relief from any provision in this Order or any decision by the Settlement Referee. This CMO may be modified only by Court order, stipulation of the parties or noticed motion (if a stipulation cannot be reached).

It is so ordered:

Dated:

Judge of the Superior Court

EXHIBIT A

CMO time line

35 days after appearance or service of CMO All non-privileged documents to be deposited

35 days after appearance or service of CMO Responses to Insurance and Scope of Work Interrogatories

30 days after Court signs CMO Plaintiffs' Preliminary Defect List due

45 days after Court signs CMO Deadline to add new Defendants and Cross-Defendants

Requests for visual inspections and non-invasive inspections due

Visual inspections and non-invasive testing

Plaintiff's Destructive Testing

Plaintiffs' Final Defect List and Cost of Repair Estimate due

Plaintiffs' Demand to Developers

Defendants' and Cross-Defendants' Requests for Destructive Testing due

Plaintiffs to publish Destructive Testing schedule

Defendants' and Cross-Defendants' Destructive Testing

Developer Demands

Expert Witness designation

Expert Meeting

Settlement Conferences

To be set by Court Depositions begin

To be set by Court Trial

EXHIBIT "B"

DESCRIPTION OF DOCUMENTS TO BE DEPOSITED INTO COURT-ORDERED DOCUMENT DEPOSITORY

A. DEFENDANTS AND CROSS-DEFENDANTS

1. Any and all non-privileged job files, building contracts, agreements, notes, correspondence, photographs, videotapes, diagrams, plans, specifications, shop drawings, "as-built" plans, calculations, journals, invoices, purchase orders, change orders, addenda reports (including reports prepared by consultants and design professionals for the original construction), job diaries, receipts, project files, site records, daily job logs, field orders, superintendent reports, requests for clarification, requests for information, time cards, governmental inspection punch lists and sign off sheets, and any and all other writings as per California Evidence Code Section 250 and invoices relating to the construction, repair, or maintenance of the Property but excluding such documents and data that were generated during pre-litigation testing pursuant to mediation between Plaintiff and Defendants (as used herein, the term "Property" refers to Plaintiff's single family residence located at _____ the land upon which such residence was constructed and all improvements appurtenant thereto).

2. All recommendations of soils and geotechnical engineers who provided services before and during construction of the Property with respect to grading, removal, backfill and compaction of the soil of the Property. All reports and requirements issued or imposed by any governmental agency having jurisdiction over the grading, removal, backfill and compaction of the soil of the Property with respect to same. All records of tests, reports, field observations, and compaction tests, of soils and geotechnical engineers who provided services before and during construction of the Property with respect to the soils of the Property. All other letters, memoranda, reports or other documents generated by soils or geotechnical engineers, governmental agencies, Defendants, consultants retained by Defendant before and during construction of the Property, or design professionals retained by Defendants before and during construction of the Property with respect to the grading, removal, backfill or compaction of the soil of the Property or any problems with same.

3. All reports, recommendations and other documents of any architect or engineer retained by Defendants before or during construction of the Property with regard to the design and construction of the post-tensioned concrete slab under the Property, including requirements articulated by any design professional with respect to the condition of the soils under such slab.

4. All Documents referring or related to Commercial General Liability or similar insurance coverage for the Property, or for you since the inception of the Property, or the Development (the term "Development" as used herein means the Dreamland Park project in which the Property is located) including, without limitation,

subcontractor insurance policies on which you are named as an additional insured, and any and all Additional Insured Endorsements and certificates which potentially provide insurance coverage for any claim asserted against each party, regardless of whether coverage has been reserved or denied by any insurance company.

5. Any and all contract proposals, contracts, subcontracts, subcontract proposals, bids, agreements, invoices, purchase orders, change orders, subcontract change orders, change order logs, addenda, diaries, and other writings related to any and all repairs and all alterations, modifications and/or improvements which relate to the Property.

6. Any and all purchase contracts, sales files, customer service files, advertisement files, marketing documents, warranty claims and warranty documents which relate to the Property.

7. All documents which embody or constitute any contract with any architect, engineer, or other design professional, and any modifications thereto, for work on the Properties.

8. All non-privileged data relating to the condition of soils at the Property, including, but not limited to manometer readings, soils sampling, soils compaction tests, geologic composition, seismic activity and seismic stability generated before and during construction of the Property but excluding such documents and data that were generated during pre-litigation testing pursuant to mediation between Plaintiff and Defendants or generated during testing and investigation conducted pursuant to this CMO prior to the time such information is required to be disclosed under the provisions of this CMO.

9. All Documents which constitute, refer or relate to Bid Proposals for provision of materials or work on the Property.

10. All logs, indices, or summaries of Documents referring or related to the Property, including without limitation, file indices, correspondence and fax logs, drawing submittal logs, and change order logs.

11. All invoices for work or materials supplied to the Property.

12. All Documents referring or related to the scheduling of work on the Property, as planned and as actually performed.

13. All Documents which constitute, refer or relate to matters causing delay or extra costs for work on the Property.

14. All non-privileged meeting notes from meetings relating to work on the Property, between your personnel and sub contractors or consultants who performed work on the Property.

15. All non-privileged documents which refer or relate to inspections of work on the Property by your representatives, any construction lender or investor, or by any public entity.

16. All plans and shop drawings for the Property.

17. All warranties for the Property or any portion thereof, whether provided by you or any subcontractor, sub-subcontractor or material supplier.

18. All non-privileged photographs, videotapes, drawings, or other graphic documentation with respect to the Property.

19. All contractors' licenses held by you from the start of construction on the Property to the present.

20. All sales brochures and advertising copy for the Property or any part thereof.

21. All correspondence and memoranda, consisting in or related to tender of defense or indemnity for the claims set forth in the present lawsuit, to any subcontractor, design professional or material supplier.
22. All Documents which refer or relate to your purchase of the land on which the Property was built, or of the Property, including without limitation, contracts, contract proposals, correspondence, deeds, and escrow instructions.
23. Any and all expert documents to be exchanged under this order.
24. Any and all documents subpoenaed from third parties as required by this order.

B. PLAINTIFF
1. Any and all non-privileged documents as defined in California Evidence Code section 250, including, but not limited to, memoranda, correspondence, invoices, repair contracts, maintenance agreements, any documents relating to costs incurred to test for and/or repair any alleged defects, contractor/repair contracts, and any other correspondence, reports, notes, memoranda, photographs and video tapes relating to any alleged defects and/or repairs, or written complaints of Plaintiff to Defendants or third parties, excluding documents and data generated from pre-litigation testing conducted pursuant to the mediation between Plaintiff and Defendants and arising from investigation conducted pursuant to this CMO.
2. Initial walk-through punch list, purchase agreements, grant deeds, warranty service claims, correspondence to or from developer and/or subcontractors regarding defects, receipts for repairs, estimate for repairs, maintenance records, appraisal documents, escrow documents, documents arising from any attempts by Plaintiff to refinance the Property that bear upon the value of the Property or an economic advantage of Plaintiff that was lost as a result of Plaintiff's inability to refinance the Property, assignment of rights, photographs or videos depicting alleged defects and any non-privileged documents relating to the value of the Property.
3. Any and all plans, maps, blueprints, or specifications relating to the construction, modification, maintenance or repair of the Property.
4. Any and all contract proposals, bids, contracts, agreements, invoices, purchase orders, change orders, addenda, accounting records, job diaries, reports, writings, and any other documents relating to any and all alterations, modifications, improvements, repair, operation, maintenance, inspections, and/or additions to the Property.
5. All documents concerning any first party insurance Property claims or warranty claims by the Plaintiff for relevant Property damage at the Property.
6. Preliminary and final defect lists and cost estimates as required by this Order.
7. Any and all expert documents as required by this Order.
8. Any and all documents subpoenaed from third parties as required by this Order.
9. All non-privileged data relating to the condition of the soils at the Property, including but not limited to, manometer readings, soils sampling, soils compaction tests, geologic composition, seismic activity and seismic stability, excluding documents and data generated during pre-litigation testing pursuant to the mediation between Plaintiff and Defendants.
10. All non-privileged photographs, videotapes, drawings, or other graphic documentation with respect to the Properties, excluding documents and data generated during pre-litigation testing pursuant to the mediation between Plaintiff and Defendants.

EXHIBIT "C"

INSURANCE AND SCOPE OF WORK INTERROGATORIES

The term "**Policy of Insurance**" refers to any agreement under which any insurance carrier(s) may be liable to satisfy the judgment.

The term "**Damages**" shall mean any actual or alleged weakness, fault, flaw, blemish, incomplete work, leak or condition causing any form of water infiltration or any construction condition indicating a failure to comply with the applicable plans or specifications, or a failure to comply with any applicable building codes, construction requirements or applicable standards in the construction industry.

The term "**Development**," means those living units and common areas comprising the residential Development known as Dreamland Park located in Hollywood, California. The term "Property" means the Plaintiff's single family residence located at 1234 Dreamland Drive and the land upon which it is built.

The terms "**you**" and "**your**" mean the responding party or your respective client and include each person and/or entity acting on its behalf, including, but not limited to, all directors, officers, and agents of the responding party.

The term "**Labor**" shall mean all work or toil. "**Laborers**" refer to workers for wages and profit including superintendents and the supervision of work.

The term "**Services**" shall mean all work done or duty performed for you, and anything useful such as supplies, installation, repairs, maintenance or acts rendered by one person to another, the former being bound to submit its will to the direction and control of the latter.

INTERROGATORY NO. 1:

Are you a corporation? If so, state:

a. the name stated in the current articles of incorporation;
b. all other names used by the corporation during the past ten (10) years and the dates each was used;
c. the date and place of incorporation;
d. the address of the principal place of business;
e. whether you are qualified to do business in California.

INTERROGATORY NO. 2:

Are you a partnership? If so, state:

a. the current partnership name;
b. all other names by the partnership during the past ten (10) years and the dates each was used;

c. whether you are a limited partnership and, if so, under the laws of what jurisdiction;

d. the name and address of each general partner;

e. the address of the principal place of business.

Interrogatory No. 3:

Are you a joint venture? If so, state:

a. the current joint venture name;

b. all other names used by the joint venture during the past ten (10) years and the dates each was used;

c. the name and address of each joint venture;

d. the address of the principal place of business.

Interrogatory No. 4:

Are you an unincorporated association? If so, state:
a. the current unincorporated association name;

b. all other names by the unincorporated association during the past ten (10) years and the dates each was used;

c. the address of the principal place of business.

Interrogatory No. 5:

Have you done business under a fictitious name during the past ten (10) years? If so, for each fictitious name state:

a. the name;

b. the dates each was used;

c. the state and county of each fictitious name filing;

d. the address of the principal place of business.

Interrogatory No. 6:

Within the past five (5) years has any public entity registered or licensed your businesses? If so, for each license or registration:

a. identify the license or registration;

b. state the name of the public entity;

c. state the dates of issuance and expiration.

Interrogatory No. 7:

At the time of the acts alleged in the Complaint, was there in effect any policy of insurance through which you were or might be insured in any manner (for example: primary, pro-rata, or excess liability coverage) for the damages, claims or actions that have arisen out of the damages at the Development? If so, for each policy state:

a. the kind of coverage;

b. the name and address of the insurance company;

c. the name, address, and telephone number of each named insured;

d. the policy number;

e. the policy period for each policy number;

f. the limits of coverage for each type of coverage contained in the policy;

g. whether any reservation of rights and/or controversy of coverage dispute exists between you and the insurance company;

h. the name, address, and telephone number of the custodian of the policy;

i. whether the policy provides broad form coverage; and

j. whether you have an Additional Insured Endorsement running in favor of any other party to this litigation.

Interrogatory No. 8:

Are you self-insured under any statute for the damages, claims or actions that have arisen out of the damages at the Development or the Property? If so, specify the statute.

Interrogatory No. 9:

Please provide the last known name, address, and telephone number of the person or persons most knowledgeable regarding the work and/or services you performed for the Development or the Property, and state whether this person or persons is currently employed by you:

a. with respect to the bidding and contracting for the work you performed on the Development or the Property;

b. with respect to the work performed at the Development or the Property

Interrogatory No. 10:

Please provide the name, address, and telephone number of the current owner, partner, managing agent, or employee who is most knowledgeable regarding the work and/or services you performed at the Development or the Property.

Interrogatory No. 11:

Identify your job foreman/construction supervisor for the Development or the Property, providing name, last known address and telephone number, and state whether this person is currently employed by you.

Interrogatory No. 12:

Provide the name, title, last known address and telephone number of the person at your company at the time you performed your work at the Development or the Property who was responsible for obtaining your insurance policies and additional insured endorsements, and state whether this person is currently employed by you.

Interrogatory No. 13:

What work did you perform on the Development or the Property (including work performed or serviced provided under verbal or written contracts, change orders and/or extras)?

Interrogatory No. 14:

Where did you perform work on the Development or the Property (by phase number, unit number, address of single family residence or building number)?

Interrogatory No. 15:

State the first and last dates you performed work on the Development or the Property.

Interrogatory No. 16:

With what person or entity did you contract to perform the above-described work?

Interrogatory No. 17:

Did you supply materials to the Development or the Property?

Interrogatory No. 18:

If you supplied materials, describe the materials you provided, and state the first and last dates on which you supplied materials.

Interrogatory No. 19:

If you supplied materials, state to what person(s) or entity (ies) you provided them, and provide their last known address and telephone number.

Interrogatory No. 20:

Did you Contract any of the work that was to be performed by you on the Development or the Property to another person or entity?

Interrogatory No. 21:

If you did Contract any of your work to another, identify the person(s) or entity (ies) to whom you Contracted, providing their last known address and telephone number.

Interrogatory No. 22:

If you did Contract any of your work to another, was that Contract in writing?

Interrogatory No. 23:

If you did Contract any of your work to another, state what work and/or services you so Contracted.

EXHBIT "D"
VISUAL INSPECTION/INVASIVE TESTING REQUEST

   I. Your Client:
  II. Work Performed:
 III. Addresses Requested for Inspection/Testing:
 IV. Type of Inspection/Testing Requested:
  V. Special Equipment Required:
 VI. Estimated Time Required:

The Mann Law Firm
Robert S. Mann, Esq., State Bar No. 77283
2029 Century Park East, 19th Floor
Los Angeles, California 90067-9998

Telephone (310) 556-1500 Fax (310) 556-1500

Attorneys for Plaintiff

Superior Court of the State of California for the County of Los Angeles

	) Case No. _____
	)
	)
Unhappy Homeowner,	) **Complaint for:**
	)
	)
Plaintiff,	) 1. Strict Liability
v.	) 2. Negligence
Mega Builders Development Co.	) 3. Breach of Implied Warranty of
(and Does 1 - 100, inclusive,)	) Fitness
Defendants.	) 4. Breach of Implied Warranty
	) of Merchantability
_____	) 5. Breach of Contract

Plaintiff alleges:

GENERAL ALLEGATIONS

1. Plaintiff is and at all times herein mentioned was an individual residing in the County of Los Angeles, State of California.

2. At all times herein mentioned, Plaintiff is and was the owner of a single family residence in a gated residential community called Dreamland Park. Plaintiff's residence is located at and commonly known as 1234 Dreamland Way the legal description of which is _____, as set forth in the Records of the County Recorder's Office of the County of Los Angeles, and shall be referred to hereinafter as the "Residence."

3. Plaintiff is informed and believes and based thereon alleges that Defendant Mega Builders Development Co. (hereinafter "Mega") is and was a California Limited Partnership, authorized to do business and doing business in the City and County of Los Angeles and State of California and is and at all times relevant herein was the developer of Dreamland Park and the Residence.

4. The true names or capacities, whether individual, corporate, associate or otherwise, of Defendants named in this action as Does 1 through 100, inclusive, are unknown to Plaintiff, who therefore sues said Defendants by such fictitious names; Plaintiff will ask leave to amend this Complaint to show their true names and capacities when same have been ascertained. Plaintiff is informed and believes, and on that basis alleges, that each of the Defendants designated as a Doe herein, is responsible in some manner for the events and happenings hereinafter described.

5. Plaintiff is informed and believes, and on that basis alleges, that at all times herein mentioned, each of the Defendants was the agent, servant, and/or employee of each of the remaining Defendants, and at all times were acting within the purpose and scope of said agency and/or employment, each with the consent of the others.

6. Plaintiff is informed and believes and based upon such information and belief alleges that Mega is and was, at all times relevant herein, a builder and developer of mass-produced properties for sale to the general public. Mega was and is the builder and developer of Dreamland Park and is and was the builder and developer of the Residence. As builder, developer and seller of the Residence, and its component parts and systems, Mega knew that the Residence would be sold to and used by members of the general public for the purpose of a residential dwelling, and Mega knew or reasonably should have known that Plaintiff, as the purchaser of the Residence, would not, in the exercise of reasonable diligence, know or should know of the defects as alleged herein.

7. In or about April, 1996, Plaintiff, on the one side, and Mega, on the other side, entered into a written agreement of purchase and sale pursuant to which Plaintiff subsequently purchased the Residence from Mega.

8. Plaintiff is informed and believes, and based thereon alleges, that Mega failed to properly design the Residence, failed to properly perform work and labor and/or failed to provide proper materials with respect to the construction of the Residence, all of which has resulted in defective and deficient work and damage to the Residence. Plaintiff has provided notice to Mega of such defective and deficient construction.

9. Plaintiff is informed and believes, and based thereon alleges, that the Residence as designed and built is defective and that such defective design and construction has resulted in the following defects and deficiencies, among others: unstable and improperly compacted soil under and around the Residence; cracks in and damage to the structural foundations; cracks in the floors, walls and ceilings; movement and rotation of the Residence causing doors and windows to become out of plumb, damage to finished surfaces, and damage to hardscape adjacent to the Residence. Unless repaired, the damage from such construction defects will continue to occur, and the Residence will continue to deteriorate and degrade in the future.

10. The foregoing description of defects is only partial, and Plaintiff will seek leave to amend this Complaint or will prepare a final list of defects describing the full nature and extent of defects once ascertained.

11. Each of the defects and deficiencies in the design, construction, labor and materials provided, performed or supplied by Mega in connection with the design and construction of the Residence were unknown to Plaintiff, are latent and not reasonably discoverable by plaintiff. Plaintiff is informed and believes, and based upon such information and belief alleges that the Residence may contain additional defects and deficiencies not presently known, but which will be alleged by way of amendment or established at the time of trial, according to proof.

12. Mega knew or had reason to know that the purchasers of homes in Dreamland Park, including Plaintiff, would rely on the skill, judgment, and expertise of Mega as developer and builder in developing and building the Residence so that the Residence would be reasonably fit for its intended purpose.

FIRST CAUSE OF ACTION

For Strict Liability for Defective Product

(As Against Mega and Does 1-25, inclusive)

13. Plaintiff realleges and incorporates herein by reference paragraphs 1 through 12, inclusive, of his Common Allegations as though set forth in full hereat.

14. Defendants, and each of them, as developers, owners, mass-producers, manufacturers of component parts and systems, builders, and sellers of the Residence are strictly liable and responsible to Plaintiff for all damage sustained as a result of the defects and deficiencies alleged herein and those that may be discovered before the trial of this action.

15. As a direct and proximate result of the conduct, acts and/or omissions by defendants, and each of them, Plaintiff has suffered damages in an amount precisely unknown, but which Plaintiff is informed and believes to be in excess of $50,000, including expenses incurred to pay professionals and other contractors to investigate and expenses to be incurred to repair the Residence and correct the defective construction. Plaintiff has suffered a loss of use and enjoyment of the Residence. In addition, the value of the Residence has been significantly and materially diminished as a result of the construction defects and deficiencies. Plaintiff will establish the precise amount of such damages at trial, according to proof.

16. As a further result of the conduct of defendants, and each of them, Plaintiff will be required to vacate the Residence and will incur moving costs and the expenses of renting living quarters pending and during repair and reconstruction of the Residence.

SECOND CAUSE OF ACTION

For Negligence

(Defective Construction Against Mega and Does 1-25, inclusive)

17. Plaintiff realleges and incorporates herein by reference paragraphs 1 through 12, inclusive, of his Common Allegations and paragraphs 13 through 16 of his First Cause of Action as though set forth in full hereat.

18. Plaintiff is informed and believes, and based thereon alleges, that defendants, and each of them, were and are developers, designers, builders, contractors, construction managers and/or other persons or entities who participated in the process of construction of the Residence and who performed works of labor and/or provided services necessary for the building and construction of the Residence. In so doing, defendants, and each of them, caused the Residence to be constructed through their own Development, construction, design, works of labor, their supplying of materials, and their equipment and services, and by performing works of Development, design, labor, and by supplying materials, equipment and services in order to complete the Residence.

19. Plaintiff is informed and believes, and based thereon alleges, that defendants, and each of them, negligently, carelessly, tortiously and wrongfully failed to use reasonable care in the Development, design, supervision, construction, installation and implementation of the various construction elements, as set forth above, with regard to the Development, design, supervision, and construction of the Residence.

20. Plaintiff is informed and believes, and based thereon alleges, that defendants, and each of them, whether acting as developer, designer, builder, contractor or otherwise, performed work, labor and/or services for or upon the Residence, knew or should have known that if the Residence was not properly or adequately constructed, that the owners and occupiers of the Residence, such as Plaintiff, would be substantially damaged thereby and that the Residence would be defective.

21. Defendants, and each of them, were under a duty to exercise ordinary care as developers, designers, builders, contractors, construction managers or otherwise to avoid reasonably foreseeable injury to Plaintiff, as the owner and occupier of the Residence, and to the Residence. Defendants, and each of them, knew or should have foreseen with reasonable certainty that Plaintiff would suffer the monetary and other damages set forth herein if defendants, and each of them, failed to perform its duty to cause the Residence to be constructed in a proper and workmanlike manner and fashion.

22. In performing the work of a developer, designer, builder, contractor, construction manager or otherwise, defendants, and each of them, failed and neglected to perform the design, work, labor and services properly or adequately; and defendants, and each of them so negligently, carelessly and in an unworkmanlike manner performed the design, work, labor and/or services that the Residence was constructed

improperly, and contains the defects described herein and such other defects as may be discovered before the trial of this action.

23. As a direct and proximate result of the conduct, acts and/or omissions by defendants, and each of them, all of which conduct, acts and/or omissions were discovered by Plaintiff within one (1) year last past and which could not have been discovered prior thereto in the exercise of reasonable diligence by Plaintiff, Plaintiff has suffered damages in an amount precisely unknown, but which Plaintiff is informed and believes to be in excess of $50,000, including expenses incurred to pay professionals and other contractors to investigate and expenses to be incurred to repair the Residence and correct the defective construction. Plaintiff has suffered a loss of use and enjoyment of the Residence. In addition, the value of the Residence has been significantly and materially diminished as a result of the construction defects and deficiencies. Plaintiff will establish the precise amount of such damages at trial, according to proof.

24. As a further result of the conduct of defendants, and each of them, Plaintiff will be required to vacate the Residence and will incur moving costs and the expenses of renting living quarters pending and during repair and reconstruction of the Residence.

THIRD CAUSE OF ACTION

For Breach of Implied Warranty of Fitness

(Against Mega and Does 1-25, inclusive)

25. Plaintiff realleges paragraphs 1 through 12, inclusive, of his Common Allegations, paragraphs13 through 16, inclusive, of his First Cause of Action, and paragraphs 17 through 24 of his Second Cause of Action, and incorporates them by reference as though set forth in full hereat.

26. Defendants, and each of them, knew that the Residence would be used as single family residence, intended for personal use and further knew or had reason to know that, among other things:
   A. The soil beneath the Residence should be prepared to support the Residence so that the Residence would not subside, rotate, lean or otherwise suffer damage;
   B. Other construction components, including the foundations, should be adequately constructed and installed to provide a proper, secured inhabitable home;
   C. The aforementioned components should be incorporated into the Residence so that Plaintiff and other inhabitants of the home would enjoy a stable, usable, secure and otherwise habitable home.

27. Defendants, and each of them, knew or had reason to know that Plaintiff, as purchaser of a single family home, would rely upon the skill and judgment of defendants, and each of them, in the construction of the Residence and in performing the construction and installation and in supervising the construction and installation of components and materials fit for their particular purposes as described herein.

28. Defendants, and each of them, impliedly warranted that the aforementioned components as constructed and installed at the Residence would be fit for the particular purposes as alleged above.

29. The Residence was not fit for said purposes but, in fact, was and is unsound and uninhabitable and has failed as alleged herein.

30. As a direct and proximate result of the conduct, acts and/or omissions by defendants, and each of them, Plaintiff has suffered damages in an amount precisely unknown, but which Plaintiff is informed and believes to be in excess of $50,000, including expenses incurred to pay professionals and other contractors to investigate and expenses to be incurred to repair the Residence and correct the defective construction. Plaintiff has suffered a loss of use and enjoyment of the Residence. In addition, the value of the Residence has been significantly and materially diminished as a result of the construction defects and deficiencies. Plaintiff will establish the precise amount of such damages at trial, according to proof.

31. As a further result of the conduct of defendants, and each of them, Plaintiff will be required to vacate the Residence and will incur moving costs and the expenses of renting living quarters pending and during repair and reconstruction of the Residence.

FOURTH CAUSE OF ACTION

For Breach of Implied Warranty of Merchantability

(Against Mega and Does 1-25, inclusive)

32. Plaintiff realleges paragraphs 1 through 12, inclusive, of his Common Allegations, paragraphs 13 through 16, inclusive, of his First Cause of Action, and paragraphs 17 through 24 of his Second Cause of Action, and incorporates them by reference as though set forth in full hereat.

33. Defendants, and each of them, at all times and places alleged herein, impliedly warranted that the Residence was of merchantable quality. The Defendants, and each of them, have breached the warranty of merchantability, by virtue of the following failures, among others:

  A. The soil beneath the Residence should be prepared to support the Residence so that the Residence would not subside, rotate, lean or otherwise suffer damage;
  B. Other construction components, including the foundations, should be adequately constructed and installed to provide a proper, secured inhabitable home;
  C. The aforementioned components should be incorporated into the Residence so that the Residence was of merchantable quality.

34. As a direct and proximate result of the conduct, acts and/or omissions by defendants, and each of them, Plaintiff has suffered damages in an amount precisely unknown, but which Plaintiff is informed and believes to be in excess of $50,000, including expenses incurred to pay professionals and other contractors to investigate and expenses to be incurred to repair the Residence and correct the defective construction. Plaintiff has suffered a loss of use and enjoyment of the Residence. In addition, the value of the Residence has been significantly and materially diminished as a result of the construction defects and deficiencies. Plaintiff will establish the precise amount of such damages at trial, according to proof.

35. As a further result of the conduct of defendants, and each of them, Plaintiff will be required to vacate the Residence and will incur moving costs and the expenses of renting living quarters pending and during repair and reconstruction of the Residence.

FIFTH CAUSE OF ACTION

For Breach of Contract

(As Against Mega and Does 26-50)

36. Plaintiff realleges paragraphs 1 through 12, inclusive, of his Common Allegations, paragraphs13 through 16, inclusive, of his First Cause of Action, and paragraphs 17 through 24 of his Second Cause of Action, and incorporates them by reference as though set forth in full hereat.

37. The purchase and sale contract (hereinafter the "Contract") stated, among other things, that the Residence was free from known defects in materials, construction and design. In consideration thereof, Plaintiff paid Mega the purchase price for the Residence.

38. Within four (4) years last past, Mega breached the terms of the agreement of purchase and sale by delivering the Residence to Plaintiff with substantial defects in construction, installation, design and materials.

39. Plaintiff has performed all obligations required of him under the Contract, save and except those obligations the performance thereof has been relieved by the breaches of the Contract by Mega, as set forth herein.

40. As a direct and proximate result of the conduct, acts and/or omissions by defendants, and each of them, Plaintiff has suffered damages in an amount precisely unknown, but which Plaintiff is informed and believes to be in excess of $50,000, including expenses incurred to pay professionals and other contractors to investigate and expenses to be incurred to repair the Residence and correct the defective construction. Plaintiff has suffered a loss of use and enjoyment of the Residence. In addition, the value of the Residence has been significantly and materially diminished as a result of the construction defects and deficiencies. Plaintiff will establish the precise amount of such damages at trial, according to proof.

41. As a further result of the conduct of defendants, and each of them, Plaintiff will be required to vacate the Residence and will incur moving costs and the expenses of renting living quarters pending and during repair and reconstruction of the Residence.

WHEREFORE, Plaintiff prays for judgment against Defendants, and each of them, as follows:

1. On the First Cause of Action for Strict Liability for compensatory damages in an amount to be determined by the trier of fact;

2. On the Second Cause of Action for Negligence for compensatory damages in an amount to be determined by the trier of fact;

3. On the Third Cause of Action for Breach of Implied Warranty of Fitness, for compensatory damages in an amount to be determined by the trier of fact;

4. On the Fourth Cause of Action for Breach of Implied Warranty of Merchantability, for compensatory damages in an amount to be determined by the trier of fact;

5. On the Fifth Cause of Action for Breach of Contract against Mega, for compensatory damages in an amount to be determined by the trier of fact;

6. For costs of suit incurred herein;

7. For professional and technical fees in amount unknown at the present time and subject to proof at time of trial;

8. For loss of use in an amount unknown at the present time; and

9. For such other and further relief as the Court deems just and proper.

Dated: _____

The Mann Law Firm
Robert S. Mann, Esq.

By: _____

Robert S. Mann

Attorneys for Plaintiff Unhappy Homeowner

The Mann Law Firm
Robert S. Mann, Esq., State Bar No. 77283
2029 Century Park East, 19th Floor
Los Angeles, California 90067-9998
Telephone (310) 556-1500

Attorneys for Plaintiff Homeowner

Superior Court of the State of California For the County of Los Angeles

	)	Case No.
	)	
	)	
	)	
	)	
Unhappy Homeowner,	)	Complaint For:
	)	
Plaintiff,	)	1. Breach of Contract
v.	)	2. Breach of Implied Warranties;
Defective Constructors,	)	3. Breach of Express Warranty;
and Does 1 - 100, inclusive	)	4. Negligence; and
Defendants.	)	5. Breach of Fiduciary Duty
	)	
	)	

Plaintiff alleges:

GENERAL ALLEGATIONS

1. Plaintiff Unhappy Homeowner is, and at all times herein mentioned was, an individual residing in the county of Los Angeles, State of California.

2. At all times herein mentioned Plaintiff was the owner of certain real Property commonly known as 100 Luxury Lane, Beverly Hills, California, which shall be referred to as "the Property."

3. Plaintiff is informed and believes and based thereon alleges that at all times herein referred to, Defendant Defective Constructors, Inc. ("Contractor"), is and at all times relevant herein was:

   A. A sole proprietorship doing business in the County of Los Angeles, State of California and wholly owned by Dan Defects;

   B. Licensed as a general contractor by the State of California.

4. The true names or capacities, whether individual, corporate, associate or otherwise, of Defendants named in this action as Does 1 through 100, inclusive, are unknown to Plaintiff, who therefore sues said Defendants by such fictitious names; Plaintiff will ask leave to amend this Complaint to show their true names and capacities when same have been ascertained. Plaintiff is informed and believes, and on that basis alleges, that each of the Defendants designated as a Doe herein, is responsible in some manner for the events and happenings hereinafter described.

5. Plaintiff is informed and believes, and on that basis alleges, that at all times herein mentioned, each of the Defendants was the agent, servant, and/or employee of each of the remaining Defendants, and at all times were acting within the purpose and scope of said agency and/or employment, each with the consent of the others.

6. On or about _____ Plaintiff and Contractor entered into a written contract (the "Contract") whereby Contractor agreed to, among other things, provide all labor, equipment, tools, transportation, materials and services for the construction of a single family home, consisting of a main residence, pool cabana, swimming pool, tennis court pavilion and tennis court (collectively the "Residence") on the Property in accordance with the plans and specifications provided to the Contractor. Under the Contract, Plaintiff was to pay for the actual cost of labor and materials for the construction, and the Contractor was to receive a fixed contractor's fee of ____ plus payment for labor and materials provided directly by the Contractor. The Contract further provided as follows:

   A. The Contractor specifically warranted and covenanted that the work to be performed and materials to be used in connection with the construction would be of good quality and free from faults and defects;

   B. The Contractor specifically covenanted that a relationship of trust and confidence existed between Contractor and Plaintiff, and that the Contractor would use his best skill and judgment and use every effort and do all things necessary to perform the work in the most expeditious and economical manner consistent with good workmanship, sound business practice and the best interests of the Plaintiff, including the negotiation and letting of subcontracts;

   C. The Contractor was authorized to retain subcontractors to perform labor and services and/or provide materials to the construction of the residence; however, the Contractor was responsible for all work and material provided by said subcontractors and was to supervise said subcontractors to ensure the quality of their work;

   D. With respect to any item of work the cost of which exceeds $20,000 for which the Contractor retained a subcontractor, the Contractor was required to obtain, in good faith, three written bids from subcontractors and submit the bids to the Plaintiff and the Plaintiff's representative on the project; the Plaintiff was to have five (5) days to select the subcontractor; however, if the Plaintiff did not make a selection within five business days, then the Contractor was permitted to select the subcontractor who submitted the lowest bid;

E. The Contractor was not to enter into any agreement or other arrangement for the furnishing of labor, materials and/or services for any of the work with an "affiliated entity," unless the Contractor fully disclosed in writing said agreement or arrangement with an affiliated entity to the Plaintiff and Plaintiff approved of the agreement or arrangement; and

F. If any litigation, arbitration or other proceedings are commenced between the parties concerning the Contract, then the prevailing party is entitled to an award of the attorneys' fees incurred in connection with such litigation, arbitration or other proceeding.

7. Thereafter, the Contractor commenced the construction of the residence purportedly in accordance with the terms and conditions of the Contract. Plaintiff is informed and believes that while Contractor retained some subcontractors to provide some labor and/or materials to the construction of the Residence, the Contractor provided much of the labor, services and materials directly to and for the Plaintiff.

8. Plaintiff is informed and believes and based thereon alleges that the Contractor failed to properly perform the work and labor and/or failed to provide proper materials, and failed to adequately supervise the subcontractors which Contractor retained, all of which has resulted in defective and deficient work. Plaintiff has provided notice to Contractor of defective and deficient construction.

9. Plaintiff is informed and believes and based thereon alleges that the Residence as built is defective and fails to comply with the plans and specifications, in the following areas, among others:

A. Improper installation of the terrace membrane waterproofing, flashing and improper sloping of the terraces, resulting in water intrusion into the framing of the Residence, causing deterioration in the framing members and resultant damage to the framing, sheathing, interior ceilings, walls and floors of the Residence;

B. Improper installation of planter boxes, including defective waterproofing and flashing, resulting in water intrusion into the interior of the Residence and causing damage to the framing, sheathing, ceilings and walls of the Residence;

C. Improper installation of the roofing and roofing flashing system, resulting in water intrusion into and damage to the framing, sheathing and the interior of the Residence;

D. Improper installation of the waterproofing membrane and flashing system at the door and window openings, resulting in water intrusion, causing damage to the framing, sheathing, ceilings, walls and floor of the Residence;

E. Improper installation of waterproofing and flashing at the walls for the tennis pavilion, resulting in water intrusion into and damage to the tennis pavilion;

F. Improper installation of waterproofing and flashing at the foundation walls of the pool cabana, resulting in water intrusion into and damage to the pool cabana; and

G. Improper application of lath and stucco, resulting in water intrusion into the Residence and damage to the exterior stucco, framing, sheathing and interior walls and finishes.

10. The foregoing list of defects is only partial, and Plaintiff will seek leave to amend this Complaint or will prepare a final list of defects describing the full nature and extent of defects once ascertained.

11. Plaintiff is informed and believes and based thereon alleges that in connection with the construction, the Contractor failed to act in the best interest of the Plaintiff as required by the Contract. Plaintiff is further informed and believes and based thereon alleges that, among other things, the Contractor:

A. Retained subcontractors to perform work, the cost of which exceeded $20,000, without first obtaining three written bids from subcontractors for said work;

B. Failed to submit said bids to the Plaintiff and the Plaintiff's representative on the project;

C. Failed to obtain competitively priced bids for the work;

D. Failed to disclose to Plaintiff that work was performed by "affiliated entities" of Contractor, as such term is defined in the Contract.

FIRST CAUSE OF ACTION

For Breach of Contract

12. Plaintiff realleges and incorporates herein by reference paragraphs 1 through 11, inclusive, of his Common Allegations as though fully set forth hereat.

13. Defendants, and each of them, breached the Contract by having, among other things:

A. Failed to properly perform the work and/or failed to provide proper materials;

B. Failed to adequately supervise the subcontractors, in accordance with the terms of the Contract;

C. Failed to act in the best interests of the Plaintiff;

D. Retained subcontractors to perform work, the cost of which exceeded $20,000, without first obtaining three written bids from subcontractors for said work;

E. Failed to submit said bids to the Plaintiff and the Plaintiff's representative on the project;

E. Failed to obtain competitively priced subcontracts; and

F. Used affiliated entities without disclosing same to Plaintiff.

14. Plaintiff has performed all conditions, covenants, and promises required by him to be performed in accordance with the terms and conditions of the agreement.

15. Plaintiff is informed and believes, and based thereon alleges, that the residence may be additionally defective in ways and to extents now precisely unknown, but which will be inserted herein by way of amendment or will be established at the time of trial, according to proof.

16. Plaintiff is informed and believes, and based thereon alleges, that as a direct and proximate result of the Defendants' breach of the Contract, and the resulting defective condition of the Residence, Plaintiff has suffered damages in an amount precisely unknown, but which Plaintiff is informed and believes to be in excess of $1 million, to pay professionals and other contractors to repair the Residence and

correct the defective construction. Plaintiff will establish the precise amount of such damages at trial, according to proof.

17. As a further result of Defendants' breach of the Contract, and the resulting defective construction of the Residence, Plaintiff will be required to vacate the Residence and will incur moving costs and the expenses of renting living quarters pending repair and reconstruction of the Residence.

18. As a proximate result of Defendants' breach of the Contract, Plaintiff has also suffered damages as a result of overcharges for cost of the construction, the amount of which is unknown to Plaintiff, but which Plaintiff is informed and believes to be in excess of $1 million. Plaintiff will establish the precise amount of such damages at trial, according to proof.

19. Under the terms of the Contract, Plaintiff is further entitled to an award of the attorneys' fees incurred in connection with this action.

SECOND CAUSE OF ACTION

For Breach of Implied Warranty of Fitness

20. Plaintiff realleges and incorporates herein by reference paragraphs 1 through 11, inclusive, of his Common Allegations as though fully set forth hereat.

21. Defendants knew that the Residence would be used as single family residence, intended for Plaintiff's personal use and further knew or had reason to know:

   A. The weather and water exposed surfaces and systems should be constructed and installed to provide a weather and water-tight barrier for the home;
   B. Other construction components including stucco, terraces and planters should be adequately constructed and installed to provide a proper, secured inhabitable home;
   C. Defendants further knew that the aforementioned components should be incorporated into the Residence so that Plaintiff and other inhabitants of the home would enjoy a water-tight, stable, usable, secure and otherwise habitable home.

22. Defendants knew or had reason to know that Plaintiff would rely upon the skill and judgment of Defendants in the construction of the Residence and in performing the construction and installation and in supervising the construction and installation of components and materials fit for their particular purposes as described herein.

23. Defendants impliedly warranted that the aforementioned components as constructed and installed at the Residence would be fit for the particular purposes as alleged above.

24. The Residence was not fit for said purposes but, in fact, was unsound and uninhabitable and failed as alleged herein.

25. As a proximate and direct result of such breaches of the implied warranty of fitness for particular purpose, plaintiff has sustained and will sustain damages in an amount precisely unknown, but which Plaintiff is informed and believes to be in excess of $1 million, to pay professionals and another contractors to repair the Residence and correct the defective construction. Plaintiff will establish the precise amount of such damages at trial, according to proof.

26. As a further result of Defendants' breach of the Contract, Plaintiff will be required to vacate the Residence and will incur moving costs and the expenses of renting living quarters pending repair and reconstruction of the Residence.

THIRD CAUSE OF ACTION

For Breach of Express Warranty

26. Plaintiff realleges and incorporates herein by reference paragraphs 1 through 11, inclusive, of his Common Allegations, and paragraphs 13 through 19, inclusive, of his First Cause of Action, as though fully set forth hereat.

27. Pursuant to the terms of the Contract, the Contractor expressly warranted to Plaintiff that the Residence would be of good and substantial quality, free from faults and defects and built in accordance with the terms of the Contract.

28. The Plaintiff contracted with the Contractor in reliance on the express warranties, affirmation of fact, and promises made by Defendants. Plaintiff has duly performed all the conditions and covenants of said Contract on his part to be performed.

29. As set forth herein, the Residence was not in good quality, free from faults and defects and not built in accordance with the terms of the Contract.

30. Plaintiff notified Defendants of said breach of warranties, and said Defendants have refused, and continue to refuse, to remedy these defects. Plaintiff intends, by service of the summons and the complaint on the Defendants, to further notify said Defendants of the facts set forth herein as to the Defendants' breach of said express warranties.

31. Plaintiff relied on the Defendants, and each of their, express representations and warranties that the Residence would be of good quality, and free from fault and defects and built in accordance with the terms of the Contract.

32. Plaintiff is informed and believes, and based thereon alleges, that as a direct and proximate result of the Defendants' breach of the express warranty, Plaintiff has suffered damages in an amount precisely unknown, but which Plaintiff is informed and believes to be in excess of $1 million, to pay professionals and other contractors to repair the Residence and correct the defective construction. Plaintiff will establish the precise amount of such damages at trial, according to proof.

33. As a further result of Defendants' breach of express warranty, Plaintiff will be required to vacate the Residence and will incur moving costs and the expenses of renting living quarters pending repair and reconstruction of the Residence.

34. Under the terms of the Contract, Plaintiff is further entitled to an award of the attorneys' fees incurred in connection with this action.

FOURTH CAUSE OF ACTION

For Negligence

35. Plaintiff realleges and incorporates herein by reference paragraphs 1 through 11, inclusive, of his Common Allegations as though fully set forth hereat.

36. Plaintiff is informed and believes, and based thereon alleges, that Defendants were and are builders, contractors, construction managers and/or other persons or entities who participated in the process of construction of the Residence and who performed works of labor and/or services necessary for the building and construction, including the supervision of construction of the Residence. In so doing, Defendants, and each of them, caused the Residence to be constructed through their own works of labor, their supplying of materials, their equipment and services, and through causing and supervising other contractors and subcontractors, including other Defendants, to perform works of labor, to supply materials, equipment and services, in order to complete the Residence.

37. Plaintiff is informed and believes, and based thereon alleges, that Defendants negligently, carelessly, tortiously and wrongfully failed to use reasonable care in the supervision, construction, installation and implementation of the numerous construction elements, as set forth above.

38. Plaintiff is informed and believes, and based thereon alleges, that the Defendants named herein, whether builder, contractor or otherwise, performed work, labor and/or services upon the Residence and each of them knew or should have known that if the Residence was not properly or adequately constructed, that Plaintiff would be substantially damaged thereby and that the Residence would be defective. Likewise, Defendants knew or reasonably should have known that if other building systems and elements were not adequately constructed or installed, that Plaintiff would be substantially damaged thereby and the Residence would be defective.

39. The Defendants were under a duty to exercise ordinary care as builder, contractor, construction manager or otherwise to avoid reasonably foreseeable injury to Plaintiff and the Residence. Defendants knew or should have foreseen with reasonable certainty that Plaintiff would suffer the monetary and other damages set forth herein if said Defendants failed to perform their duty to cause the Residence to be constructed in a proper and workmanlike manner and fashion.

40. In performing the works of a builder, contractor, construction manager or otherwise, the Defendants failed and neglected to perform the work, labor and services properly or adequately; each Defendant so negligently, carelessly and in an unworkmanlike manner performed the work, labor and/or services that the Residence was constructed improperly, and contains the defects described herein.

41. As a direct and proximate result of the foregoing negligence, carelessness and unworkmanlike conduct, action and/or omissions by Defendants, Plaintiff has suffered damages in an amount precisely unknown, but which Plaintiff is informed and believes to be in excess of $1 million, to pay professionals and other contractors to repair the Residence and correct the defective construction. Plaintiff will establish the precise amount of such damages at trial, according to proof.

42. As a further result of Defendants' negligence, Plaintiff will be required to vacate the Residence and will incur moving costs and the expenses of renting living quarters pending repair and reconstruction of the Residence.

FIFTH CAUSE OF ACTION

For Breach of Fiduciary Duty

43. Plaintiff realleges and incorporates herein by reference paragraphs 1 through 11, inclusive, of his Common Allegations as though fully set forth hereat.

44. By virtue of the provisions of the Contract, at all times relevant herein, Defendants, and each of them, owed to Plaintiff a fiduciary duty of trust and confidence, so that Defendants, and each of them, were required at all times during the course of construction to act on behalf of the best interests of Plaintiff.

45. Such duty of trust and confidence required that, among other things, Defendants, and each of them, were to ensure that the Residence was built in the most expeditious and economical manner, to obtain competitive bids from subcontractors for work in excess of $20,000 and, in the event the Plaintiff did not select a subcontractor from those competitive bids and to select the subcontractor who had submitted the lowest bid.

46. Plaintiff did, in fact, repose trust and confidence in Defendants, and each of them, and reasonably relied upon the Defendants' skill and judgment and reasonably expected that Defendants would act in Plaintiff's best interests. Such reliance was reasonable by virtue of the fact that Defendants, and each of them, expressly undertook such trust and confidence pursuant to the terms of the Contract.

47. Plaintiff is informed and believes and based thereon alleges that Defendants, and each of them, breached their fiduciary obligations to Plaintiff by virtue of the following, among other things:

    A. Defendants failed to perform the work in the most expeditious and economic manner in the best interests of the Plaintiff;
    B. Defendants retained subcontractors to perform work, the cost of which exceeded $20,000, without first obtaining three written bids from subcontractors for said work;
    C. Defendants failed to submit said bids to the Plaintiff and the Plaintiff's representative on the project; and
    D. Defendants failed to obtain competitive bids for subcontracts.

48. As a proximate result of Defendants' breach of fiduciary duty as set forth herein, Plaintiff has suffered damages as a result of overcharges for cost of the construction, the amount of which is unknown to Plaintiff, but which Plaintiff is informed and believes to be in excess of $1 million. Plaintiff will establish the precise amount of such damages at trial, according to proof.

49. The Defendants, and each of them, in breaching their fiduciary obligations to Plaintiff, acted with oppression, fraud and malice within the purview of Civil Code Section3294. Accordingly, Plaintiff is entitled to an award of punitive damages against said Defendants, and each of them, in a sum to be determined at the time of trial.

WHEREFORE, Plaintiff prays for judgment against Defendants, and each of them, jointly and severally, as follows:

1. On the First Cause of Action for Breach of Contract:

    A. For compensatory damages in excess of $1 million;

   B. For attorneys' fees incurred herein;

2. On the Second Cause of Action for Breach of Implied Warranty for compensatory damages in excess of $1 million;

3. On the Third Cause of Action for Breach of Express Warranty:
   A. For compensatory damages in excess of $1 million;
   B. For attorneys' fees incurred herein;

4. On the Fourth Cause of Action for Negligence, for compensatory damages in excess of $1 million;

5. On the Fifth Cause of Action for Breach of Fiduciary Duty:
   A. For compensatory damages in excess of $1 million;
   B. For punitive damages in an amount to be determined by the trier of fact;

6. For costs of suit incurred herein;

7. For professional and technical fees in amount unknown at the present time and subject to proof at time of trial;

8. For loss of use in an amount unknown at the present time; and

9. For such other and further relief as the Court deems just and proper.

Dated: _____,

                        The Mann Law Firm,
                        Robert S. Mann, Esq.

By: _____

                        Robert S. Mann

                        Attorneys for Plaintiff
                        Unhappy Homeowner

The Mann Law Firm
Robert S. Mann, Esq., State Bar No. 77283
2029 Century Park East, 19th Floor
Los Angeles, California 90067
Telephone (310) 556-1500
Facsimile (310) 556-1577

Attorneys for Plaintiff Unhappy Homeowner

Superior Court of the State of California For the County of Los Angeles

	)	Case No.
	)	
	)	
	)	
	)	
Unhappy Homeowner,	)	**Final Defect List and**
		Cost of Repair Estimate
Plaintiff,	)	
v.	)	
Mega Builders Development Co.	)	
and Does 1 to 1000, inclusive		
Defendants.	)	
_____	)	
And Related Cross Actions	)	
_____	)	

FINAL DEFECT LIST

1. Concrete/Masonry
   1.1 Front steps.
       Nature of Defect: Front steps formed incorrectly.
       Remedial Repair: Saw cut front concrete step and rebar. Demo existing concrete. Regrade and reslope front area and driveway. Remake stone skirts and repour new front steps.
   1.2 Concrete formed incorrectly at back steps at columns. Columns, when finished, would not be supported by concrete flatwork.
       Nature of Defect: Concrete formed incorrectly at back steps adjacent to columns, insufficient space between column base and edge of flatwork.
       Remedial Repair: Demo existing concrete, form new concrete, dowel rebar to new concrete to provide base for columns.
   1.3 Improper Venting of Below Grade Masonry Walls.
       Nature of Defect: Below grade masonry walls constructed without adequate subfloor venting. Subfloor vents installed too low. Subfloor vents covered with landscape materials. CO Sensor omitted.
       Remedial Repair: Cut new openings in masonry walls for subfloor vents. Subfloor vents still too low and cannot be repaired at north side of west elevation.
   1.4 Concrete Flatwork covered discharge pipe.
       Nature of Defect: Concrete flatwork covered discharge pipe from sump pump.
       Remedial Repair: Demo concrete, repair and replace pipe, pour new concrete flatwork.
   1.5 Guesthouse Slab improperly installed.
       Nature of Defect: Slab on grade installation improperly constructed without proper embedment of 6×6 reinforcing wire, leading to extensive cracking and differential settlement of 1/8 inch.
       Partial Remedial Repair: Saw cut damaged areas, where visible, demo concrete, and replace with appropriately reinforced concrete. Other areas not yet repaired.
   1.6 Bottom of drive at entry to underground garage not properly sloped. Remedial repair (not yet made): Remove and repour concrete.
   1.7 Intentionally omitted.
   1.8 Grout not properly installed at site walls. Remedial repair: Install grout at necessary inside site walls.

2. Framing/Finish Carpentry.
   2.1 Door and Floor Heights improper.
       Nature of Defect: Framer incorrectly placed a 2×4 under flashing at exterior doors. Exterior doors were then measured and cut to incorrect height. As a result, interior floor heights were incorrect. Plans call for installation of 1 $^1/_2$ inch lightweight concrete on floors. Floor heights not framed to accommodate this installation, leaving door thresholds, stairs and fireplaces at incorrect levels. No allowance at front entry for finished floor height at front door pan and threshold. Floor at guest bedroom bath on first floor framed at improper height.
       Remedial Repair: Remove all exterior doors, remove and replace waterproofing, customize door thresholds, customize stone thresholds, elevate subfloor in interior to adjust height of floor, change height of fireplaces at fireboxes,

adjust and customize height of stair risers, change subfloor at second floor from 1 and 2 inch lightweight concrete to 3/4 inch gypcrete and adjust base-boards, reframe the main step at stairs. Remove toilet flange at first floor guest bedroom bath and reset at proper height. Reframe steps at main stairs.

2.2 Skylight improperly framed.

Nature of Defect: In bathroom No. 3, skylight improperly framed, reducing amount of light from skylight.

Remedial Repair: Install mirrored Plexiglas sheathing inside light shaft.

2.3 Intentionally omitted.

2.4 Laundry Chute improperly framed.

Nature of Defect: Laundry chute not framed so that laundry will fall into catch basin.

Remedial Repair: Customize the laundry chute with new framing and install corrective sheet metal work.

2.5 Door Jambs at Elevator improperly installed. Framing not square.

Nature of Defect: Door jambs at elevator entrance on all floors improperly installed.

Partial Remedial Repair: Install additional thresholds to repair jambs. Modify inside of elevator doors to accommodate 4 inch clearance from cabinet, install panel at top of all openings to cover gap between top of door opening and cab-inet, install L-metal, install drywall, repaint as needed, repair and paint dry-wall. Further repair: Reframe, provide new doors, install drywall, repaint.

2.6 Front Elevation not framed to plans.

Nature of Defect: Front elevation not framed to plans.

Partial Remedial Repair: Prepare new architectural drawings for new front eleva-tion details. Remove quoins, install new stucco and quoins and patch as necessary.

2.7 Quoins improperly aligned at front elevation. See above.

2.8 Weatherstripping omitted or improperly installed at exterior doors. Repair: install interlocking weatherstripping and astragals at exterior doors.

2.9 Front door sill pan not deep enough to allow door to close. Remove, fabricate new pan and install.

2.10 Planter ledge at guesthouse bar reverse sloped and not properly waterproofed. Partial Remedial repair: Remove planter ledge. Further repair: Replace planter ledge.

2.11 Large cornice over front door reversed sloped and failure to construct drainage and roofing system. Remedial repair: Remove and replace cornice and install drainage system.

2.12 Door opening to library to have door. Door opening framed and finished with-out door jamb. Repair: Remove drywall, reframe and install door jamb.

2.13 Exit door from basement to have molding. Door opening framed and finished so that molding could not be installed. One hour fire rated door jamb not installed.

2.14 Door opening in master bedroom, guesthouse, and bath No. 2 improperly framed so that molding could not be installed.

2.15 Arches at skylight in master not framed correctly and do not match.

2.16 Door opening at laundry chute not installed.

2.17 Skylight at upstairs bedroom number 2 bath (pink marble finishes) improperly framed, resulting in irregular appearance. Repair: Float and repair drywall.

2.18 Guesthouse to have spaces between exterior pre-cast moldings. Improperly framed, resulting in condition where no space could be left between moldings.

2.19 Attic access opening at roof not framed to receive waterproofing and no waterproofing installed.

2.20 Drywall screws used to secure skylights on roof. Installer should have used galvanized or stainless steel screws with neoprene washers.

2.21 Correct framing defects at breakfast room radius panels, guesthouse doors, guest room casing, dining room doors, family room doors and bonus room doors.

3. Decks/Roofs

3.1 Decks improperly installed.

Nature of Defect: Decks improperly constructed by improper sloping, lack of blocked drainage devices and improper flashing, and insufficient height at door thresholds. Drainage devices used primary drain device only instead of drainage device with secondary drainage system.

Remedial Repair: Remove existing deck surface and subsurface to framed structure. Replace with lightweight concrete properly sloped, replace all deck drains, remove and replace quoins, remove and replace flashing at exterior doors leading to decks, remove and replace diato flashing, apply elastomeric liquid polyurethane with silica sand at balconies and window sheet metal pans, repair sheet metal at windows, install secondary drainage mechanism (scupper), install curb cap, apply new waterproofing membrane, install new stucco. Remedial repair not yet done: remove and replace all balcony decks, remove structure below so provide sufficient clearance to properly slope decks to drain and retain adequate height at door thresholds. Reinstall subfloor, waterproofing membrane, drain assemblies and finished surface.

3.2 Roofs.

Nature of Defect: Entire flat roof defectively installed, not sloped, leading to water intrusion at second floor hallway. Vertical seams of roofing not properly lapped. Standing water on roof. Membrane is one torch applied layer over one base sheet. No reinforcing layer at waterways. Crickets to deck not smoothly transitioned, resulting in dissimilar height of cricket and deck. Cricket edges not beveled or chamfered. Water ponding at cricket waterways. Roof material no properly lapped over valley metal.

Remedial Repair: Remove and replace entire flat roof, correct improper slopes, reset drains, demolish stucco and sheet metal, install new sheet metal, including ridge caps, parapet coping, copper reglet and counter, GI vaulter, inlaid gutter with coping detail, repair existing copper attic vents, remove and replace quoins, path stucco at chimneys and condenser wells, cut membrane at bridge areas along walls, install 3 × 3 inch perlite cant strips, nail cut membrane to deck, torch apply 12 inch wide strip of smooth membrane over cant and cut areas, remove and replace electrical box for condensers and conduits at roof location. Install crickets as necessary, reslope roof to drain, install copper cap flashing to slope to drain.

3.3 Roof drains not installed per plans. Plans require drains to be set in sumps. Drains set on roof deck.

3.4 Galvanized flashing used instead of copper flashing as specified in plans without owner's consent.

3.5 Plans require installation of metal flashings and storm collars. Storm collars missing or improperly installed at roof mounted vents. Vents flashed using torch applied material, with open edges allowing water intrusion. Torch down application without proper heating of substrate leaves voids and inadequate adhesion, permitting water intrusion.

3.6 Vent at west chimney chase not properly installed. Vent not sealed around edges to stucco, resulting in water intrusion at chimney chase interior.

3.7 Through-wall scuppers not installed per plans or omitted.

3.8 Perimeter wall coping at junction of chimney chase walls left open, with exposed builder paper and wood, resulting in path for water intrusion into chase and interior.

3.9 Wall coping at intersecting wall left open, with exposed wood and paper. No membrane, metal transition or other moisture protection device, leaving path for moisture intrusion. Coping turned up to stucco wall, allowing water to migrate into wall cavity. Wood staining observed on top of wall.

3.10 Top of perimeter walls not waterproofed. Plans require installation of Bituthene 4000 membrane at top of walls over roof membrane. Omitted in installation.

3.11 Deck holes for PVC vents too large. No metal flashing or storm collars as per plan requirements installed. Membrane flashing substituted. Membrane flashing observed to be cracking.

3.12 Vent for heater missing top cap and storm collar and metal flashing. Vent sealed with roof cement.

3.13 Electrical box sealed at top edge with Bituthene. Improper application of Bituthene due to exposure to UV light, causing degradation of product. Repair: remove existing material and plaster, install new ice and water shield, sheet metal cap and new plaster.

3.14 Perimeter wall copings not installed properly. No expansion joints. Copings installed with simple lap joints, rivets and solder.

3.15 At north chimney chase, Z metal at chase sides extends behind the pre-cast stone corner, resulting in condition where water runs behind the cast stone and must dissipate through the stone.

3.16 Improperly large gaps between chimney stacks and chase covers, resulting in path for water intrusion.

3.17 At west chimney, galvanized chase top in direct contact with copper chase cover.

3.18 Deteriorated sealant at chase cover where set into kerf cut in cast stone, resulting in path for water intrusion.

3.19 Slate roof underlayment buckled.

3.20 Mechanical equipment platforms not installed on underlying deck of roof.

3.21 Window dormers not properly attached to roof. Fasteners working out of roof underlayment. Plywood roof underlayment found to be deteriorated at edges where fasteners installed.

3.22 Required UL Class A fire rating classification cannot be achieved with specified roof membrane if installed directly onto plywood substrate. Installation required barrier board over plywood.

4. Moisture Proofing

4.1 Roof Equipment improperly waterproofed.

Nature of Defect: Penetrations for electrical and refrigerant lines improperly waterproofed.

Remedial Repair: Remove stucco and lath, install new waterproofing material, install new lath and exterior plaster with urethane sealant.

4.2 Windows and doors.

Nature of Defect: Window moisture proofing installed defectively. Specifically, flashing, sill pans, lathing were all installed defectively, leading to moisture intrusion at windows. Sill pans lacked vertical legs or were installed with vertical legs shorter than those specified or required by SMACNA or industry standards and practices; sill pans had vertical flanges bent to horizontal or otherwise installed improperly. Sill pans not sealed or soldered at corners. Door sill pans too low for finished floor heights. Main front door sill pan too narrow for front door. Bituthene-type self-adhering membrane separated from window frames. Bituthene-type self-adhering membrane reversed lapped. Saw kerf at bottom of pre-cast stone specified but not installed. Bituthene material at windows and doors at guesthouse degraded due to exposure to UV. Failure to follow manufacturer's recommendations to limit direct exposure to sunlight to no more than 60 days. In some locations at guesthouse, moistop paper installed in lieu of proper ice and water shield at doors and windows, in violation of plans and specifications.

Remedial Repair: Remove all pre-cast from exterior, remove all doors and windows, remove interior moldings and casings, customize thresholds for height errors, add additional slope underneath sill pans at doors, apply elastomeric coating with silica sand at sill pans, install new flashing, sill pans, add additional slope under sill pans, install ice and water shield waterproofing, install diamond lath as necessary, modify secondary sill pans at oval windows, modify oval window head flashings, provide kerf cut at window and doors frames, fabricate and install sheet metal flashing system for windows and doors, reinstall and adjust windows and doors, seal all seams, install Z-bar flashing and shed flashing at window above front entry, reinstall moldings, casings, thresholds. Install new pre-cast. Substantial pre-cast omitted from reinstallation, still to be installed.

4.3 Exterior doors. See 4.2 above for nature of defect and remedial repair.

4.4 Below Grade waterproofing.

Nature of Defect: Below grade waterproofing at NE wall at basement, adjacent to media room failed.

Remedial Repair: Excavate earth from exterior wall to a depth of 10 feet. Repair damaged below grade waterproofing, install modified weep screed and reglet flashing at perimeter of house, replace exposed waterproofing and tie into existing waterproofing below grade, install bituthene membrane, apply Mira Dry waterproofing membrane above weep screed and lap six inches onto existing framing, backfill and compact soil under direction of soil engineer, remove and replace drywall and insulation in media room, drain water from media room, eliminate moisture from media room, remove scaffolding used during repair work. Smell at basement media room remains.

4.5 Below Grade waterproofing at Front Elevation.

Nature of Defect: Below grade waterproofing exposed when soil removed from front elevation.

Remedial Repair: Install stainless steel cover on front of wall.

4.6 Lathing. See Stucco, below.

4.7 Tennis Court retaining wall subdrain improperly installed.
Nature of Defect: Tennis court retaining wall subdrain not sloped to drain.
Remedial Repair Recommendation: Excavate and reinstall subdrain to slope properly to drain.

4.8 Small balconies at front elevation not waterproofed. Remedial Repair: Install waterproofing.

4.9 Perforated drain at north and east elevations, adjacent to formal dining room not properly sloped to drain. Remedial Repair (not yet made): Remove soil from above drain, remove and replace drain, replace below grade waterproofing as necessary, backfill and replace landscaping as necessary.

4.10 Below grade waterproofing at south elevation between garden gate and stairs exposed to weather and deteriorated. Remedial Repair: Remove and replace.

4.11 Second floor small balcony over front entry defectively installed and installed in variance from plans. Plans called for installation of copper roof. Stucco nailed to bottom of gutter. Diamond lath and double layer of building paper under stucco found to be rusted and wet. Repair: Remove existing stucco, install weep screed at top edge of gutter with extended flange and reinstall stucco.

4.12 Copper gutter at small balcony over front entry defectively installed. Gutter installed directly to wood framing, without builder paper between gutter and framing as per SMACNA recommendations and good practice. Repair: Remove gutter, install waterproof membrane and reinstall gutter.

4.13 Plans required installation of metal flashings behind cast stone at headers of windows and doors. Omitted in installation, resulting in path for moisture migration into wood framing and interior.

4.14 At second floor ledge, sheet metal flashing behind cast stone installed with open mitered corners providing path for water intrusion.

4.15 Z metal flashing at radius wall at roof eave open at stucco wall, with exposed wood backer, providing path for water intrusion into interior.

5. Mechanical

5.1 HVAC. Plans specified that outlet and return air grilles be centered on walls as appropriate. Grilles installed improperly without centering on walls. Remedial Repair (not completed): Relocate grilles. Remedial Repair: revise moldings as necessary to fit around improperly located outlets and return air.

5.1.1 Air conditioning units throughout left uncovered and exposed, requiring cleaning, including washing scraping, replacement of deteriorated fiberglass and filters and removal of dirt from ductwork.

5.1.2 HVAC lines dented by drywall installer. Repair: Remove and replace as necessary.

5.2 Plumbing.

5.2.1 Noise.

Nature of Defect: ABS pipe installed in violation of specifications in plans calling for cast iron, improper installation of pipe without soundproofing as required by plans and code.

Remedial Repair: Perform inspection by acoustical engineer, remove and replace multiple runs of ABS pipe with cast iron, removal of drywall to expose pipe, replace, patch and paint drywall, correct framing as necessary, insulate pipes, install quiet flush toilets, install soundproofing insulation flanges under all toilets. Noise at library remains.

5.2.2 Debris in water lines.

Nature of Defect: Substantial amount of dirt and small rocks in water lines and plumbing fixtures, causing malfunction and premature wear of fixtures. Pressure lost at fixtures.

Remedial Repair: Backflush plumbing lines. Replace fixtures as necessary. Has not repaired problem. Remedial Repair still under investigation.

5.2.3 Intentionally omitted.

5.2.4 Main 3 inch gas line left exposed to weather and rusted. City inspector required that line be replaced.

Nature of Defect: 3 inch main gas line was exposed during construction and rusted.

Remedial Repair: Replace line.

5.2.5 Plumbing and drain, waste and vent lines installed outside of framing, intruding into living spaces (Note: this defect may also be attributable to framing and finish carpentry categories).

Remedial Repair: Cover with soffits, drywall, framing, or cabinets as appropriate. Aesthetic appearance problem not corrected.

5.2.5 Gas Valves.

Nature of defect: Incorrect check valves installed, no earthquake shut-off valve installed.

Remedial Repair: Relocate gas pipes and install correct valves.

5.2.6 Sump pump inoperable.

Nature of Defect: Sump pump installed at basement level does not operate. Discharge line blocked. Sump not formed with sufficient space for two sump pumps.

Remedial Repair: Reform sump, repair and replace discharge line, install secondary sump pump. Work in progress.

5.2.7 Drain pipe improperly installed in her master closet.

Nature of Defect: Drain line not constructed in wall framing, intrudes into closet.

Remedial Repair: Install chamfered soffit detail to conceal exposed drain line.

5.2.8 No plumbing/wiring for Mr. Steam unit. Repair: Remove stone work, install plumbing and wiring, fabricate new stone work and install.

5.2.9 Laundry room drain improperly installed. Design and fabricate custom cover to camouflage defective drain installation.

5.2.10 Gas lines to fireplaces in Living room and Family room improperly installed, requiring relocation to below floor installation.

5.2.11 Wiring and plumbing for Mr. Steam units in her master bath not installed. Repair: Remove stone work, install wiring and plumbing lines, purchase and install new stone work for bath.

5.2.12 Connection from fire department connection to fire sprinkler riser not installed. Repair: Install line from Fire Department Connection to fire sprinkler riser.

6. Electrical
  6.1 Service Panel. Service panel covers purchased but lost on site by contractor. Repair: Purchase and install new panels. Shell only installed at 600 amp main panel. Repair: Install necessary components in service panel shell.
  6.2 Fixtures improperly aligned in interior. Repair: Relocate junction boxes.
  6.3 Exterior junction boxes for sconces improperly aligned. Repair: Relocate junction boxes.
  6.4 Circuit breaker and panels paid for but missing from job. Repair: Purchase and install new breakers and panels. See 6.1, above.
  6.5 Interior switches, junction boxes and light cans improperly installed. Repair: Relocate junction boxes, rewire defectively installed switches, create a one-hour fire rated enclosure for basement due to incorrect installation of boxes, relocate and rewire wiring in exhaust hood, add switch at 4 gang switch, install variable speed control switch for kitchen exhaust fan, install missing switches, and relocate outlets and switches in kitchen; uncover recessed fixtures in basement where covered with drywall.
  6.6 Rough electrical for outside wall sconce on wall at south side of exterior door to study omitted. Rough electrical for other outside wall fixtures not centered and installed too far behind finished exterior wall surface.
  6.7 Electrical boxes for receptacles improperly installed at varying heights throughout the home and not within code specifications with regard to set back from finished wall surface. Repair: relocate and reinstall in proper locations with proper depth from face of finished wall surface.
  6.8 Galvanized electrical conduit in various locations installed against copper piping.
  6.9 Copper and foam-encased copper conduit lines placed through and against galvanized wall flashings at roof. Conduits not sealed at wall penetrations. Flexible conduit placed through walls at lower west roof. Code requires rigid conduit or adequate flashing to prevent water intrusion.
  6.10 Relocate two switches adjacent to door jambs not installed as per plans.
  6.11 Improper size electrical conduit used in run to main electrical panel.
  6.12 Electrical panels purchased by client but missing. Purchase and install electrical panels.
  6.13 Create a one hour fire rated wall for recessed fixtures at basement due to lack of fire rated installation.
  6.14 Owner purchased Mr. Steam unit. Contractor lost or gave away unit. Owner required to purchase second unit.
  6.15 Electrical circuits at kitchen of guesthouse omitted. Repair: Install circuits.
  6.16 Sconce at bathroom of bedroom no. 1 improperly located. Repair: Relocate sconce (repair not yet made).
7. Stucco
  7.1 Scratch Coat improperly installed.
     Nature of Defect: Stucco scratch coat defectively installed, leading to a failure of the brown coat to adhere. Scratch coat not scored to depth required by code and industry practices, lath not secured to building, scratch coat too smooth for adhesion of brown coat.
     Remedial Repair: Remove existing stucco coat and lath. Install new lath, and stucco. Demolish stucco at corbels, waterproof and replace with new stucco. Demolish stucco at library chimney, apply waterproofing and restucco. Demolish

stucco at ledge detail mid-way up chimney, apply cant strip, apply waterproofing, restucco. Add adhesive admix to brown coat.

7.2 Wrought Iron penetrations improperly lathed.

Nature of Defect: Wrought iron penetrations at 36 locations improperly lathed leading to water intrusion.

Remedial Repair: Remove lath and exterior plaster. Reapply lath and scratch coat, apply liquid adhesive and ice and water shield at wrought iron areas, apply brown and finish coat of exterior plaster.

7.2 Lathing improper around windows and doors. See above.

7.3 Lathing not secured to building.

Nature of Defect: Lathing not secured to building surface.

Remedial Repair: remove and replace lath and exterior stucco. See above.

7.4 Weep screed improperly installed.

Nature of Defect: Weep screed improperly installed or omitted at main house wall in area adjacent to garage and at guesthouse.

Remedial Repair: Remove and replace weep screed and exterior waterproofing.

7.5 Stucco improperly installed at Chimney.

Nature of Defect: Stucco and lath installed without adequate waterproofing.

Remedial Repair: Remove stucco and lath, install ice and water shield, demolish the ledge detail, install cant strip and install new lath and stucco.

7.6 Stucco improperly installed at corbels and quoins. See above, No. 7.5

7.7 Stucco installed too low in entry outside front door: interfered with installation of limestone pavers.

7.8 Stucco installed below grade on south side of driveway. Remedial repair: Cut stucco to above grade and install weep screed as necessary.

7.9 Quoins improperly installed, sealing omitted at juncture of quoin and exterior plaster relief joint. No drainage provided along plaster relief joint, allowing water to stand in relief joint. Quoin attachment method deviated from that specified on plans. Plans specified attachment with concealed epoxied bolts and metal straps. Actual installation used screws driven through builder paper, with no waterproofing measures.

7.10 Roof edge flashings not properly installed at transition to vertical wall surfaces. Remedial Repair: Remove and replace.

7.11 Exterior window sills at guesthouse embedded in stucco. Stucco or window should have been installed so that window sill would butt against finished stucco surface.

7.12 Stucco covered outlet box for exterior sconce at front elevation.

8. Drywall

8.1 Nail Pops. Drywall improperly secured to framing, leading to nail pops throughout home. Problem continues to manifest itself.

8.2 Drywall improperly secured to walls and ceilings. Drywall not nailed and screwed at edges as per applicable building codes and industry standards, leading to drywall nail pops and loose attachment of drywall. Repair: Renail and secure drywall and float as necessary to correct defects.

8.3 Drywall not flat in plane. In numerous locations throughout the home, drywall surface was not flat in plane, leading to gaps at door frames, moldings and finishes, including baseboards and crown moldings. Repair: Float as necessary to attain uniform flatness.

8.4 Improper thickness of drywall installed at side of doors at breakfast area. Repair: Remove and replace with thinner drywall.

8.5 Radius arched ceiling at first floor improperly installed resulting in irregular curve. Repair by patching and floating as necessary.

8.6 Corner of radius wall at second floor incorrectly framed and drywalled. Repair: Remove drywall, demolish framing and correct framing, reinstall drywall and finish.

8.7 Drywall surfaces required patching at various locations. Patch, float and finish as necessary.

8.8 Vents at ceiling in basement require drywall to achieve one-hour fire separation. It is believed that no drywall was installed on the outside of the vent assemblies. Repair: Remove and install drywall to achieve one-hour fire separation.

8.9 Install tape and mud at ceiling recessed fixtures.

8.10 Drywall damaged at HVAC installation, requiring replacement of four drywall sheets.

9. Miscellaneous

9.1 Chimney flue loose and insecure. Repair: Resecure chimney flue, provide and install fireplace cap, pipe, and storm drain collar and strap flue in chimney per City of Beverly Hills inspector.

9.2 Intentionally omitted.

9.3 Expansion joint omitted at cold joint where CMU chimney chase and wood frame structure of chimney chase transition.

9.4 Failure to protect copper dormers during construction resulted in damage to all copper dormers, requiring removal and replacement.

9.5 Guesthouse constructed in wrong location.

9.6 Ventilation grilles in basement blocked by fiberglass insulation.

9.7 Power ventilation system in basement omitted.

9.8 Various windows in guesthouse not primed prior to installation and allowed to weather, resulting in formation of mildew on window surfaces and warping and deterioration of window assemblies.

9.9 Vent at kitchen of guesthouse improperly installed by setting vent directly on scratch coat, lath and building paper. Vent should have been installed on framing members.

COST OF REPAIR ESTIMATE

10. Cost of Repair

10.1 General note re cost of repair. Some repairs have been completed. Some repairs have been partially completed. Other repairs are in process and some repairs have not yet commenced.

10.2 Repairs which have been partially or completely made.

10.2.1 Concrete/Masonry: $20,601.00

10.2.2 Framing/Finish Carpentry $12,662.00

10.2.3 Decks/Roofs $89,357.00

10.2.4 Moisture Proofing $324,492.00

10.2.5 Mechanical $49,721.00

10.2.6 Electrical $19,779.00

10.2.7 Stucco $10,542.00

10.2.8 Drywall $18,402.00

10.2.9 Miscellaneous (includes scaffolding expense) $13,914.00

10.2.10 Balcony decks $15,000.00

Subtotal: $574,470.00

10.3 Repairs which have not been made

10.3.1 Perforated drain $9,500.00

10.3.2 Dirt in plumbing lines $4,000.00

10.3.3 Repair of sump pump in basement. Work in progress—time and materials.

10.3.4 Install pre-cast at doors and windows $146,320.00

10.4 Expert/Architectural Fees Continuing

10.5 Attorneys Fees Continuing

10.6 Increased General Conditions due to corrective work: $141,528.00

10.7 Overcharges by contractor for work not done: $27,800.00

10.8 Relocation Expense $76,800.00

10.9 Moving and Storage $3,000.00

10.10 Additional Construction Loan Interest Continuing

10.11 Additional General Conditions for Work to be Done Continuing

10.12 Telephone and Utilities Continuing

Subtotal: $408,948.00

Estimated total: $983,418.00

The Mann Law Firm
Robert S. Mann, Esq., State Bar No. 77283
2029 Century Park East, 19th Floor
Los Angeles, California 90067-9998
Telephone (310) 556-1500
Fax (310) 556-1500

Attorneys for Plaintiff Unhappy Homeowner

Superior Court of the State of California For the County of Los Angeles

	)	Case No.
Unhappy Homeowner,	)	Assigned to: Honorable ____
Plaintiff,	)	**Confidential Mediation/Msc**
	)	**Brief of Plaintiff Unhappy Homeowner**
	)	
v.	)	Date: May 26 and 27, 2004
	)	Time: 9:00 a.m.
	)	Dept.:
Mega Builders Development	)	Complaint filed: December 20, 2002
(and Does 1–100, inclusive,)	)	
Defendants.	)	Trial Date: Sept. 15, 2004
_____	)	

To:

The Honorable_____, Judge of the above-entitled Court and to _____, Mediator: Plaintiff Unhappy Homeowner respectfully submits the following Mediation/MSC Brief.

The Mann Law Firm

Robert S. Mann, Esq.

By: _____

Robert S. Mann

Attorneys for Plaintiff Homeowner

I

FACTUAL BACKGROUND

The central issue in the resolution of this case is the method, scope and cost of repair. Liability in this case cannot be reasonably contested.

Simply stated, the plaintiff proposes to use the conventional, tried and true, and Department of Building and Safety approved, method of installing caissons and grade beams and placing a new foundation on top of the caissons and grade beams to create a stable foundation for the home. The defendant developer proposes to use a much more esoteric and controversial repair process, compaction grouting, a process that is not approved by the Department of Building and Safety and a process that will not work given the soil conditions at the subject residence.

Accordingly, this brief is devoted to an explanation of why the defendant's proposed method of repair is not suitable and why the parties should move beyond a discussion of grout injection to the costs of a caisson and grade beam repair and focus their efforts on gathering the funds to settle the case.

In order to fully understand and appreciate why the grout injection method of repair is not suitable and is unrealistic, some background information on the Development of the project, some basics of soil mechanics and some basics on the process of grout injection are necessary. Plaintiff has attempted to describe those basics in plain English with the technical terms kept to a minimum.

II

ANALYSIS OF FLAWS IN GROUT INJECTION REPAIR PROPOSAL FOR SOILS SUBSIDENCE AT THE SUBJECT RESIDENCE

1. NATURE OF THE SOILS PROBLEM

The subject home is built in what was originally a canyon. In order to create buildable flat pads, soil was moved from other locations on the site and placed in the canyon. The layer of fill soil under the Property is approximately zero feet deep at the shallowest point, and approximately 60 feet deep at the deepest point, which is the east corner of the Property. The layer of fill soil directly below the foundation of the house is approximately 5 feet at the shallowest point and 45 feet at the deepest point. Testing has disclosed that the fill was not properly compacted. As a result, the fill is gradually settling downward and outward. The foundation and superstructure of the home, because it rests on the soil, is following the downward movement of the soil. The foundation has cracked, the walls, ceilings and other surfaces have cracked, and the window and door frames have racked. The roof truss system has also begun to fail.

2. CODE AND ENGINEERING REQUIREMENTS FOR COMPACTION OF FILL SOILS

The Building Code and requirements of the soils engineers mandate the soil be compacted to at least 90 percent relative compaction. The Building Code also specifies that fill soil designed to support structures is to be free of organic material. Relative compaction is determined by using a sample of the same type of soil and determining through laboratory tests the "maximum dry density" for that type of soil. Soil that is compacted to 90 percent relative compaction is measured against the maximum dry density. Thus, properly compacted soil will be compacted to 90 percent or more of the theoretical maximum dry density for that type of soil. The relative compaction is determined by field testing. Field and laboratory testing have revealed that the existing fill soil beneath the subject lot has a degree of relative compaction as low as 73 percent and contains intermixed organic material.

3. OVERVIEW OF SOIL MECHANICS

Soil is comprised of very small irregular fragments of sand, rock, silt, or clay. The spaces between the small particles of soil are called voids. When soil is compacted, the small particles of sand, rock, silt or clay are forced closer together, making the voids smaller.

Because the particles are irregularly shaped, they cannot be forced together during the compaction process without a lubricating material to allow them to slide together. Water is used as the lubricating material. Too much water will adversely affect the compaction and too little water will also result in less than achievable compaction. The right amount of water is called "optimum moisture content." Generally, the optimum moisture content is about 50% saturation. When the soil reaches optimum moisture content, it is compacted using heavy machines which press down on the soil in layers or "lifts."

Even after the soil has been compacted to 90% or better relative compaction, there will still be some voids in the soil. As a result, the soil will still have some ability to absorb water. If the soil is comprised of very fine particles, the water will tend to stay in the voids. If the particles are larger, the water will tend to drain away.

If the soil is properly compacted, the least amount of void space will be present and the soil will not absorb excessive amounts of water, whether from irrigation, rainfall or otherwise. Hence, even large amounts of water introduced onto properly compacted soil will not cause the soil to settle.

However, if the soil is not properly compacted, like the soil beneath the subject Property, the amount of void space is much larger than the void space in properly compacted fill. As a result, water is able to permeate through the soil. Recall that the water acts as a lubricant. When the water permeates into the poorly compacted fill, it allows the particles to rearrange themselves, causing settlement. Under no circumstances, however, will the process of water permeating into existing poorly compacted fill result in a gradual compaction to 90% or anything close to it without the use of mechanical processes (heavy equipment).

4. OVERVIEW OF REPAIR METHODS

Four methods of repairing the fill soils have been suggested:

A. Large holes can be drilled into bedrock, under and around the foundation. Each hole will then be filled with a "cage" of reinforcing steel and concrete (caissons) and connected to horizontal concrete beams (grade beams). A conventional (4 to 5 inch) or post-tensioned foundation would then be built on top of the grade beams. Plaintiff has proposed this method of repair;

B. Caissons could be installed and connected directly to a thick foundation (12 to 18 inches) called a "mat foundation." This eliminates the need for grade beams. Plaintiff endorses this method of repair as well;

C. A cementitious slurry could be injected into the ground to "densify" the fill soil (grout injection). The developers have proposed this method of repair;

D. Grout could be injected into the lowest portion of the fill and caissons sunk into the densified soil. The developers have not proposed this method of repair.

Generally speaking, for new construction and for the repair of failing fill soils, caissons and grade beams, or caissons and mat foundations are the accepted "tried and true" method of repair. The use of caissons founded in bedrock eliminates the possibility of further settlement adversely affecting a residence because a foundation supported by caissons is not affected by settlement. Conversely, grout injection is sometimes suggested because it is a less expensive alternative. It is also a less satisfactory and more esoteric alternative. In many instances, re-grouting is necessary, and it may take 5 to 7 years to determine whether the grout injection has been successful.

5. IMPACT UPON THE RESIDENCE FROM ALL PROPOSED METHODS OF REPAIR

Each method of repair requires substantial damage to the interior of the residence and the exterior flatwork and landscape. In each method of repair, holes would drilled in a "grid" under and around the foundation. If grout injection were used, the holes would be about 4 inches in diameter, and would be placed about every 5 feet. Thus, in excess of 150 four-inch holes would be drilled through the foundation for grout injection. If caissons are used, the holes would be approximately 24 to 36 inches in diameter, but there would be substantially fewer holes. With either repair, all of the floor finishes would be destroyed and significant damage would be done to the finished surfaces of the walls and ceilings. The flatwork and much of the landscape would be damaged beyond repair and would require replacement.

6. REPLACEMENT OR REPAIR OF THE POST-TENSIONED SLAB

The subject residence utilizes a post-tensioned slab. In this method of construction, before the concrete slab is poured, a number of high strength steel cables are laid in a grid across the area where the slab will be located. The cables are wrapped in a plastic sheath with a lubricant between the sheath and the cable. The ends of the cables protrude beyond the edges of the slab. The concrete slab is then poured. A small steel cone is place on the cable, just outside of the edge of the slab.

A hydraulic device is then attached to the end of the cable, and the cable is stretched outward. The lubrication between the plastic sheath and the steel cable permits the cable to move freely inside the sheath, which is anchored in the concrete. A small amount of tension is subsequently released, which allows the cable to pull inwards. The cone, which is clamped to the cable, is pulled against the edge of the slab. The engineering concept is to place the entire slab in compression (caused by the tension on the cable), so that the entire slab moves as one monolithic unit to resist tilting or downward movement.

In this case, the post tensioned slab has cracked in half. In order to repair the cracked slab, or to drill holes through the slab for any of the four proposed methods of repair, the following must be done:

A. The cables must be located. The cable must then be de-tensioned by attaching a hydraulic device, stretching the cables, removing the metal cone and allowing the cables to slowly pull inwards until there is no tension remaining. The slab cannot be cut or drilled until the cables are de-tensioned because if the cables are cut while under tension they will explosively break apart and injure anyone standing nearby;

B. All cracks or holes must be saw cut out of the slab, reinforcing steel must be doweled and epoxied into the opening and new concrete poured in the saw cut opening;

C. The cables must be re-tensioned.

The existing slab has a plastic sheet vapor barrier to prevent the transmission of water vapor from the soil through the slab into the residence. The vapor barrier will be destroyed when the holes are drilled through the slab. It is impossible to install a new vapor barrier or otherwise protect the residence from moisture without replacing the slab.

7. CONCEPTUAL DESCRIPTION OF GROUT INJECTION

In its most basic form, the concept of grout injection is that a cementitious slurry (sand, cement and water) is forced into the soil. This slurry presses on the adjacent soil, forcing the particles of soil together to "densify" the surrounding soil.

The process starts by coring a four-inch hole through the slab. A jack-hammer type machine then drives three-inch pipe in four foot sections down into the soil to the desired depth (the depth of "refusal" i.e., bedrock, where the pipe cannot be driven deeper). The first section of pipe has a plug at the end. After the stack of pipe is pounded to the point of refusal, the entire stack is pulled up about one foot. A pressure hose containing the slurry is connected to the top of the stack of pipes. The grout is injected down through the stack of pipes. The grout pushes the plug out and the grout is then pushed against the surrounding soil by the high pressure hose.

The high pressure hose causes the grout to form a shape at the end of the pipe roughly like that of an upside-down lightbulb. This "bulb" of grout presses on the surrounding soil. The grout does not infiltrate into the voids of the surrounding soil, but the pressure of the grout "bulb" against the surrounding soil pushes the soil particles more closely together. If the surrounding soil is not saturated and if it is not well-compacted, the void spaces in the soil surrounding the grout bulb will be reduced, making the surrounding soil denser.

If, however, there is water in the void spaces in the area surrounding the grout bulb, the pressure of the grout will not force the water out of the voids and the surrounding soil will not be densified. Instead, the size of the grout bulb will simply be reduced and the grout process will have not accomplished its purpose.

After the lowest-most portion of soil is injected, the stack of pipe is then pulled up about three feet and the process is repeated. This process of grouting and pulling the pipe up three feet at a time continues until the topmost portion of the soil that is intended to be grouted has been grouted. The entire stack of pipe is then pulled out of the ground, a new hole is then cored 5 feet away and the process starts over again until holes have been cored and pipes driven in a grid over the entire area.

Central to the grouting process is the concept of "overburden." Overburden is the weight of the soil and any structure above the soil (such as a house and foundation). If the amount of the overburden pressure is less than the pressure exerted by the grouting equipment, the grout will not densify the surrounding soil. Instead, the grout will lift the ground above, or ooze out along the side of the pipe and travel upward, or both.

In the upper 20 feet of soil, the overburden pressure is substantially reduced (because there is less soil pressing down on the area where the grout is being injected). To compensate, the pressure of the grout equipment and the amount of grout is reduced. However, because there is less pressure and less grout, there is also less densification. After a point, the injection of grout does nothing.

8. SOILS THAT CANNOT BE SUCCESSFULLY DENSIFIED BY GROUT INJECTION

As noted above, soil that is saturated with water cannot be densified by grout injection. Grout cannot be injected with sufficient pressure to displace the water from the void spaces. When grout is introduced into saturated soil, the soil is simply moved from one place to another, but it is not densified.

Soil that contains organic material cannot be densified by grout injection. The difficulty is that the organic material itself will continue to decompose, because it is not rock, silt, sand or clay. As it decomposes, it will create new spaces and the soil around it will compress into those spaces, leading to settlement of the soil above.

9. ASSUMPTIONS MADE BY THE DEFENSE REGARDING GROUT INJECTION

In proposing grout injection, the defense has made at least three critical assumptions:

A. That the soil is suitable for grout injection;

B. That the settlement of the fill soils will not continue. The defense has made this assumption because the defense has suggested that a layer of grout be installed at the lowest level of the fill. This would leave a blanket of poorly compacted fill at least 20 feet deep over the "densified soil." If the settlement of the admittedly poorly compacted fill in the topmost 20 feet has not stopped, it will continue to settle in an uneven fashion, causing further damage to the newly repaired residence; and

C. That any continuing compression of the non-densified fill will not be "differential." In differential settlement, one part of the soil settles more than other parts. The defense acknowledges that the non-densified upper portion of the poorly compacted fill above the newly densified grout injected soil will continue to settle, perhaps as much as an inch. The defense is assuming, however, that the entire blanket of non-densified fill will settle in exactly the same manner, so that the impact upon the residence will not be noticeable. If this assumption is wrong, and there is differential settlement, the house will continue to suffer damage because some areas will tilt or rotate, in much the same manner as is occurring now.

10. CONDITIONS THAT MAKE GROUT INJECTION INFEASIBLE AT THE SUBJECT RESIDENCE

The following conditions make it unrealistic to use grout injection at the subject residence:

A. The soil is saturated

As noted above, soil that is saturated with water cannot be densified by grout injection because the grout will not displace the water from the void spaces.

B. The soil contains large amounts of organic material

As noted above, soil that contains organic material cannot be densified because the organic material will continue to decompose, leaving voids and subjecting the soil to further settlement;

C. Grout injection will have an effect on the neighboring properties

Grout injection is extremely hard to control. If the grout that is injected into the soil does densify the surrounding soil, it will either move the soil sideways or upwards.

D. Grout injection will require the destruction of the existing foundation slab

As noted above, grout injection will require approximately 150 four-inch holes to be drilled in the slab. This will, effectively, destroy the post-tensioned slab and will require the complete demolition of the slab and its replacement;

E. Grout injection cannot densify all of the soil

In order to eliminate the possibility of continued compression, all of the soil under and around the residence would have to be densified (or a system of caissons and grade beams installed that would function independent of continued settlement). The uppermost 25 feet of soil cannot be densified by grout injection, no matter what soil conditions exist. The uppermost layer of soil cannot be densified by grout injection because the overburden pressure is not sufficient to contain the grout in the soil.

11. THE GROUT INJECTION PORTION OF THE REPAIR IS APPROXIMATELY THE SAME COST AS THE INSTALLATION OF CAISSONS AND GRADE BEAMS

It will cost approximately $350,000 to perform the grout injection. Plaintiff has independently verified this cost with two grout injection contractors. The cost to install caissons and grade beams is approximately $450,000. These prices do not include the damage to or repair of the superstructure or foundation and other utilities, landscape, hardscape, relocation etc. Those costs are roughly equal for all four proposed methods of repair.

12. THE GROUT INJECTION REPAIR WILL TAKE AT LEAST SIX MONTHS: APPROXIMATELY THE SAME TIME AS THE INSTALLATION OF CAISSONS AND GRADE BEAMS

A substantial amount of the cost of the repair for the subject residence is the cost for relocation while the work is on-going. The defense estimates that it will take approximately 6 months for the grout injection repair. Assuming that the six month period does not include repair to the house itself and the surrounding area, Plaintiff believes that the installation of caissons and grade beams will take approximately the same amount of time. There is no substantial savings in using grout injection.

13. **GROUT INJECTION IS NOT AN APPROVED METHOD OF REPAIR ACCORDING TO THE CITY OF LOS ANGELES, AND THE CITY WILL REQUIRE THE OWNER TO RECORD AN AFFIDAVIT THAT MAKES HIS PROPERTY UNMARKETABLE**

The City of Los Angeles does not recognize grout injection as an approved method of repair because history has shown that this method of repair does not always work. We have verified by a direct conversation with _____, the Chief of the Grading Division of the Department of Building & Safety, that the City of Los Angeles will require the Plaintiff to record an affidavit that states that the grout injection is not approved, and that the Plaintiff or his successor, will remain liable for subsequent damage to the Property, including successive injections of grout if the initial installation of grout fails to remedy the problem. The recordation of such an affidavit renders the Property unmarketable. No affidavit is required for caissons and grade beams.

14. **IT IS UNLIKELY THAT ANY ENGINEER OR CONTRACTOR WILL GUARANTEE THE GROUT INJECTION**

A grout injection process will have to be engineered by a soils engineer and installed by a grout injection contractor. The soils engineer of record will be responsible for any damage that may occur in the future if the grout injection method of repair does not perform as perfectly as suggested by the defense. It is unlikely that a competent soils engineer will accept the potential of as much as $2 million in liability in return for the professional fees for such a design. It is similarly unlikely that a grout injection contractor will guarantee the work for the next 50 years, the useful life of a custom built home.

15. **IF GROUT INJECTION IS USED, THE DISCLOSURES REQUIRED OF PLAINTIFF WILL RENDER HIS PROPERTY UNMARKETABLE AND WORTHLESS**

As noted above, the effectiveness of the grout injection suggested repair may not be known for 5 or more years. Because the grout injection method often does not work, the Plaintiff would have to make extensive disclosures. In effect, anyone buying the home would have to assume the risk that major settlement might continue in the future, resulting in repairs in the millions of dollars. Those disclosures effectively make the residence unmarketable.

16. **THE DEFENSE HAS NO DOWNSIDE TO A GROUT INJECTION FIX AND ALL OF THE RISK HAS BEEN TRANSFERRED TO THE PROPERTY OWNER.**

Unless the defendant developer is willing to warranty, guarantee or otherwise stand behind their proposed fix for the next 50 years, the defense has no downside to their cheaper and untried method of repair. Conversely, the Plaintiff has now assumed all of the risk. If the Plaintiff settles for an amount of money that only enables him to perform grout injection, he bears the entire risk that the repair may fail, in which case he has no recourse and his Property is worthless.

III

CONCLUSION

Grout injection is not suitable for the repair of the residence. The soil conditions do not permit effective grout injection. The assumptions made with regard to continuing compression being uniform are not objective and not realistic given the recent movement of the soil. The entire notion of grout injection is much too uncertain and speculative and would require disclosures that would made the Property unmarketable.

It is worthwhile to consider what would happen if the Plaintiff had no house on his lot and he applied for a building permit. The City of Los Angeles would require him to install caissons and grade beams, as they do in every similar circumstance. They would do so because they know from experience that the installation of caissons and grade beams is the only way to insure that a house built upon poor soil will perform well over time.

There is no reason why the repair of the house should not meet the same standards. The Plaintiff did not buy a house so that he would have to worry forever whether it will crack, settle or slide down the hill. He bought an expensive house that he thought would last for many years. Had the defendant built a proper house, the Plaintiff would have gotten what he paid for. The defendant built a defective house and they must pay a sufficient amount of money to have it repaired in a proper and reasonable manner.

DATED: _____

The Mann Law Firm
Robert S. Mann, Esq.

By: _____

Robert S. Mann

Attorneys for Plaintiff Homeowner

The Mann Law Firm
Robert S. Mann, Esq..,
State Bar No. 77283
2029 Century Park East, 19th Floor
Los Angeles, CA 90067
Telephone: (310) 556-1500
Facsimile: (310) 556-1577

Attorneys for Defendants,

Superior Court of the State of California For the County of Los Angeles

)	Case No.
)	
Defective Construction, Inc.)	Complaint Filed: 4-4-00
and Dan Builder, an individual.)	
Unhappy Homeowner,)	**Answer of Defective Construction, Inc. and Dan Builder, an individual to Complaint.**
Plaintiff,)	Trial Date: None Set
)	Mcod: None Set
)	Dcod: None Set
)	
v.)	
Defective Construction, Inc.;)	
Dan Builder, an individual and)	
Does 1 through 50, inclusive,)	
Defendants)	
_____)	

And related cross-actions

Comes now Defendants Defective Construction, Inc. and Dan Builder, an individual, and in answer to the unverified Complaint herein, for themselves alone, admit, deny and allege as follows:

1. Whenever a sum of money appears in the Complaint, these answering Defendants deny said sum, and the whole thereof, or any other sum; whenever a date appears, these answering Defendants deny said date or any other date; whenever a time appears, these answering Defendants deny said time or any other time; and whenever a rate or an amount of interest appears, these answering Defendants deny said rate and amount or any other rate and amount.

2. By virtue of the provisions of Code of Civil Procedure section 431.30(d), these answering Defendants deny, generally and specifically, each and all of the allegations contained in the unverified Complaint herein, and specifically deny that Plaintiffs are entitled to any relief sought in the Complaint.

First Affirmative Defense

(Failure to State a Cause of Action)

3. The Complaint, and each and every purported cause of action set forth therein, fails to state facts sufficient to constitute a cause of action against these answering Defendants.

Second Affirmative Defense

(Statute of Limitations)

4. The Complaint, and each and every purported cause of action set forth therein, is barred by the applicable Statute of Limitations as set forth in Code of Civil Procedure 337.1, 337.15, 338, 339 and 340.

Third Affirmative Defense

(Uncertainty)

5. The Complaint, and each and every purported cause of action set forth therein, is uncertain pursuant to Code of Civil Procedure 430.10(f), and therefore fails to state a cause of action against these answering Defendants.

Fourth Affirmative Defense

(Laches)

6. Plaintiffs are barred from recovery against these answering Defendants under the doctrine of Laches.

Fifth Affirmative Defense

(Unclean Hands)

7. Plaintiffs are barred from recovery against these answering Defendants under the doctrine of Unclean Hands.

Sixth Affirmative Defense

(Waiver)

8. Plaintiffs have waived any right to recovery under the Complaint through their own affirmative misconduct and/or failure(s) to act.

Seventh Affirmative Defense

(Estoppel)

9. Plaintiffs are estopped from asserting the claims alleged in the Complaint as a result of their own affirmative misconduct and/or failure(s) to act.

Eighth Affirmative Defense

(Failure to Mitigate)

10. Plaintiffs are not entitled to an award of damages against these answering Defendants on the grounds that Plaintiffs have failed to take reasonable steps to mitigate their damages, if any.

Ninth Affirmative Defense

(Counter-Claim and Set-Off)

11. To the extent that these answering Defendants may become liable to Plaintiffs by virtue of the Complaint herein (which liability is generally and specifically denied), these answering Defendants allege that they entitled to a counter-claim and set-off as against Plaintiffs, by virtue of the facts, acts, events, breaches, transactions and occurrences alleged and referred to in the Cross-Complaint filed concurrently herein by these answering Defendants and Cross-Complainants. Said Cross-Complaint, and any and all amended or supplemental pleadings thereto, and each and every of their allegations, are hereby incorporated herein by this reference as though set forth in full hereat.

Tenth Affirmative Defense

(Comparative Negligence)

12. These answering Defendants are informed and believe and on such information and belief alleges, that the injury and damage, if any, alleged in the Complaint occurred and was proximately caused by either the sole or the partial negligence of Plaintiffs, which negligence bars or reduces Plaintiffs' recovery herein.

Eleventh Affirmative Defense

(Reduction to Percent of Fault)

13. The right of Plaintiffs to recovery herein, if any right exists, is reduced and limited to the percentage of negligence attributable to these answering Defendants pursuant to section 1431.2 of the California Civil Code.

Twelfth Affirmative Defense

(Negligence of Others)

14. These answering Defendants deny that Plaintiffs were damaged as a proximate result of any conduct on the part of these answering Defendants. These answering Defendants affirmatively alleges that Plaintiffs' damages, if any, were proximately caused by the independent conduct of third parties or entities whether or not parties to this action. Plaintiffs' recovery against these answering Defendants, if any, must therefore be reduced to the extent that those damages, if any were caused by independent conduct of third parties.

Thirteenth Affirmative Defense

(Active-Passive Negligence Defense)

15. If these answering Defendants are found responsible in damages to Plaintiffs or some other party, whether as alleged or otherwise, then Defendants are informed and believe and, on that basis allege, that the liability will be predicated upon the

active conduct of Plaintiffs or other third parties, whether by negligence, breach of warranty, strict liability in tort or otherwise, which active conduct proximately caused the alleged damage and that Plaintiffs' action against Defendants is barred by that active and affirmative conduct.

Fourteenth Affirmative Defense

(Reservation)

16. These answering Defendants presently have insufficient knowledge or information on which to form a belief as to whether they may have additional, as yet, unstated affirmative defenses available. These answering Defendants reserves herein the right to assert additional defenses in the event that the discovery indicates they would be appropriate.

Fifteenth Affirmative Defense

(Failure to Notify)

17. These answering Defendants are informed and believe, and based thereon allege, that if any defects or inadequacies exist in the work performed by these answering Defendants, which Defendants deny, Plaintiffs failed to timely notify Defendants of such conditions and failed to give these Defendants timely opportunity to remedy such conditions. This conduct by Plaintiffs bars them from any relief from these answering Defendants herein.

Wherefore, these answering Defendants prays for judgment against Plaintiffs, as follows:

1. That Plaintiffs take nothing by way of their unverified Complaint;

2. That these answering Defendants and Plaintiffs be adjudged entitled to a counter-claim and set-off on the Complaint herein, according to proof;

3. For costs of suit herein; and

4. For such other and further relief as the Court deems just and proper.

Dated: _____ .   The Mann Law Firm

By: _____

Robert S. Mann

Attorneys for Defendants Defective Construction, Inc.

The Mann Law Firm
Robert S. Mann, Esq.,
State Bar No. 77283
2029 Century Park East, 19th Floor
Los Angeles, CA 90067
Telephone: (310) 556-1500
Facsimile: (310) 556-1577

Attorneys for Defendants and Cross-Complainants,

Superior Court of the State of California For the County of Los Angeles

	)	Case No.
Defective Construction, Inc.	)	Complaint Filed: 4-4-00
Unhappy Homeowners,	)	Cross Complaint of
Plaintiffs,	)	Defective Construction, Inc., for
v.	)	1. Equitable Indemnity
Defective Construction, Inc.	)	2. Implied Indemnity
and Does 1 through 50, inclusive,	)	3. Contribution
Defendants.	)	4. Declaratory Relief
Defective Construction, Inc.	)	5. Breach of Contract
Cross-Complainants,	)	6. Work, Labor and Services Rendered
v.	)	7. Promissory Fraud
Plumbing Subcontractor No. 1;	)	

Framing Subcontractor No. 2, and	)	Trial Date:	None Set
Roes 1 through 100, inclusive.	)	Mcod:	None Set
	)	Dcod:	None Set
	)		
Cross-Defendants.	)		
_____	)		

Comes now Cross-Complainant Defective Construction, Inc. and for causes of action against Cross-Defendants and each of them, alleges as follows:

1. At all times herein mentioned, Cross-Complainant Defective Construction, Inc. was and is a California Corporation, authorized to do business and doing business in the State of California as a licensed general contractor

2. Cross-complainants are informed and believe and thereupon allege that:
   a Cross-Defendant Plumbing Subcontractor No. 1 is a California Corporation, authorized to do business and doing business in the State of California as a licensed general contractor;
   b. Cross-Defendant Framing Subcontractor No. 2 is a California Corporation, authorized to do business and doing business in the State of California as a licensed general contractor.

3. The true names and capacities of Cross-Defendants Roes 1 through 100 are unknown to Cross-Complainant. Cross-Complainant therefore sues these Cross-Defendants by such fictitious names. Cross-Complainant will seek leave to amend this Cross-Complaint to insert the true names and capacities of the fictitiously named Cross-Defendants when they have been ascertained. Each Cross-Defendant designated as ROE is responsible in some manner for the acts, occurrences and liabilities herein alleged. ROES 1 through 100 are individuals and/or entities whose negligence or other conduct, in some manner, caused or contributed to the harm allegedly sustained by Plaintiff so as to give rise to duty to indemnify Cross-Complainant.

4. Cross-Complainant is informed and believes and based thereon alleges that at all times herein, each of the Cross-Defendants were the agents, principals, subcontractors, partners, co-developers, joint venturers, and/or employees of each of the remaining Cross-Defendants and were, at all times herein mentioned, acting within the scope of such agency and employment. Cross-Complainant is informed and believes and based thereon alleges that Cross-Defendants, and each of them, and Roes 1 through 100 inclusive, at all times herein were, either individuals, sole partnerships, partnerships, and as such, were in some manner responsible for the injuries and damages claimed by Plaintiff.

5. Pursuant to the terms of a written agreement between Plaintiffs, on the one side, and Cross-Complainant, on the other side, Cross-Complainant was retained to perform construction remodeling work at the Plaintiffs' residence (hereinafter the "Residence").

6. Plaintiffs have commenced an action for among other things, negligence and breach of contract, pertaining to alleged defects and/or damages in construction work at the Residence (hereinafter the "Main Action"). Plaintiffs' complaint in the Main Action is incorporated herein by reference as though set forth in full hereat.

7. Cross-Complainant is a defendant in the Main Action, wherein the Plaintiffs claim general and special damages as may be proven at the time of trial.

FIRST CAUSE OF ACTION

(Equitable Indemnity as Against All Cross-Defendants except Unhappy Homeowners)

8. Cross-Complainant realleges each and every allegation contained in paragraphs 1 through 7, inclusive, and incorporates them as if fully set forth herein.

9. Cross-Complainant denies that its actions were improper in any way whatsoever, and knows of no act, omission or negligence on their part which was the proximate cause of the injuries alleged in this action, but if Cross-Complainant is held liable to Plaintiffs, it will be as a proximate result of primary and active negligence or other wrongful conduct of the Cross-Defendants, or one or more of them. Therefore, any liability of Cross-Complainant will be imputed on the basis of vicarious or secondary liability, and not as a result of any active negligence or other acts on the part of the Cross-Complainant. If Cross-Complainant is found liable to Plaintiffs, then said liability will be the result of the acts and omissions, whether negligent or otherwise, of each of the Cross-Defendants hereto, and that as a result thereof, Cross-Complainant is entitled to total and complete indemnity from each of the Cross-Defendants herein, to the full extent permitted by applicable law.

10. In the event that Cross-Complainant is held liable to Plaintiffs, are found liable for any sum, incurs any expense in the defense of this action, or makes any settlement with any party to this lawsuit, Cross-Complainant is entitled to be indemnified by said Cross-Defendants, and each of them, pursuant to principles set forth in American Motorcycle v. Superior Court (1978) 20 Cal.3d 578, and is entitled to judgment against Cross-Defendants, and each of them, for all sums incurred, including reasonable attorneys' fees and all costs, expenses and damages incurred by it in this litigation.

11. Cross-Complainant has demanded and continue to demand that Cross-Defendants defend and indemnify Cross-Complainant pursuant to California Code of Civil Procedure '1021.6.

SECOND CAUSE OF ACTION

(Implied Indemnity as Against All Cross-Defendants Except Unhappy Homeowners)

12. Cross-Complainant realleges each and every allegation contained in paragraphs 1 through 11, inclusive, and incorporates them as if fully set forth herein.

13. Cross-Complainant has denied, and continues to deny, any liability arising out of Plaintiffs' Complaint and any and all cross-complaints of the other parties to this action. If, however, it is found that Cross-Complainant is responsible under the law to Plaintiffs and any other Cross-Complainants, then Cross-Complainant is informed and believes and based thereon alleges that the negligence, breach of contract, and acts and/or omissions of Cross-Defendants herein, in whole or in part, contributed to the happening of the incidents alleged in Plaintiffs Complaint and caused the damage therein alleged.

14. Cross-Complainant hereby demands that Cross-Defendants, and each of them, indemnify Cross-Complainant for all damages or losses which Cross-Complainant has incurred or may incur as a result of the allegations and demands made by Cross-Complainant.

15. By reason of the foregoing, if Plaintiffs recover against this Cross-Complainant, then Cross-Complainant is entitled to total implied indemnity from these Cross-Defendants, and each of them, for damages sustained by Cross-Complainant, if any, for any sums paid by way of settlement, or in the alternative, judgment rendered against Cross-Complainant in the underlying action.

THIRD CAUSE OF ACTION

(Contribution as Against All Cross-defendants Except Unhappy Homeowners)

16. Cross-Complainant realleges each and every allegation contained in paragraphs 1 through 15, inclusive, and incorporates them as if fully set forth herein.

17. In the event that Cross-Complainant is found liable for Plaintiffs' damages, which liability is expressly denied, Cross-Complainant is entitled to contribution from Cross-Defendants, and each of them, of that portion of the judgment attributable to the percentage of comparative fault or negligence assessed or assessable against Cross-Defendant, and each of them.

FOURTH CAUSE OF ACTION

(Declaratory Relief as Against All Cross-defendants Except Unhappy Homeowners)

18. Cross-Complainant realleges each and every allegation contained in paragraphs 1 through 17, inclusive, and incorporates them as if fully set forth herein.

19. A dispute has arisen and an actual controversy now exists between Cross-Complainant and Cross-Defendants, and each of them, in that Cross-Complainant contends that it is entitled to indemnity from the Cross-Defendants, and each of them, which Cross-Defendants deny.

20. Cross-Complainant desires a judicial determination of the respective rights and duties of Cross-Complainant and Cross-Defendants, and each of them, and a further declaration that Cross-Complainant is entitled to indemnification, or in the absence of indemnification, partial indemnification from said Cross-Defendants, so that any liability to persons or entities may be apportioned among the various responsible persons or entities according to their proportionate degree of responsibility.

FIFTH CAUSE OF ACTION

(Breach of Contract Against Unhappy Homeowners and ROES)

21. Cross-Complainant realleges each and every allegation contained in paragraphs 1 through 20, inclusive, and incorporates them as if fully set forth herein.

22. On or about _____, Cross-Complainant entered into a written contract with Unhappy Homeowners by the terms of which Cross-Complainant was to provide general contracting services for the construction and remodeling of the Residence.

23. Cross-Complainant has performed all of their obligations under the contract, except for those which Cross-Defendants have waived or prevented performance.

24. Cross-Defendants have failed to pay for services rendered by Cross-Complainant pursuant to the contract and are thereby in breach thereof.

25. As a proximate result of Cross-Defendants' breach as herein alleged, Cross-Complainant has suffered damages in an amount to be established at the time of trial but in excess of the sum of 50,000.00.

SIXTH CAUSE OF ACTION

(Work, Labor, and Services Rendered Against Unhappy Homeowners and ROES)

26. Cross-Complainant realleges each and every allegation contained in paragraphs 1 through 25, inclusive, and incorporates them as if fully set forth herein.

27. Within two years last past, Cross-Defendants became indebted to Cross-Complainant in the agreed amount of $75,320,00. for work, labor, and services rendered by Cross-Complainant to Cross-Defendants at the special request of Cross-Defendants.

28. No part of this amount have been paid though demand for payment in full has been made, and there is now due, owing, and unpaid from Cross-defendant to Cross-Complainant the amount of $75,320.00, together with interest at the rate of ten (10%) per cent per annum.

SEVENTH CAUSE OF ACTION

(Promissory Fraud Against Unhappy Homeowners and Roes)

29. Cross-Complainant realleges each and every allegation contained in paragraphs 1 through 28, inclusive, and incorporates them as if fully set forth herein.

30. In order to induce Cross-Complainant to continue to provide labor and materials for the Residence, Cross-Defendants represented that they would timely pay for labor and materials provided by Cross-Complainant pursuant to the Contract.

31. These representations were, in fact, false. The true facts were that Cross-Defendants never at any time intended to comply with these representations, to perform under the terms of the Contract so long as Cross-Defendants could manufacture some purported grounds for refusing to compensate Cross-Complainant. The true facts were also that Cross-Defendants intended to withhold the benefits of the Contract on unsubstantiated and non-meritorious grounds and made these representations with reckless disregard for their truth or falsity, and thereby made said representations with the intent to defraud Cross-Complainant.

32. Cross-Complainant, at the time these representations were made by Cross-Defendants, was ignorant of the falsity of the representations and believed them to be true. Cross-Complainant justifiably relied upon said promises to its detriment, and believing that Cross-Defendants would compensate Cross-Complainant, Cross-Complainant performed all of their obligations and rendered services under the contract.

33. As a proximate result of Cross-Defendants' conduct, Cross-Complainant has suffered, and will continue to suffer in the future, damages, interest and attorneys fees in a sum to be established a the time of trial.

34. Cross-Defendants' conduct described herein was intended by the Cross-Defendants to cause injury to Cross-Complainant or was despicable conduct carried on by the Cross-Defendants with a willful and conscious disregard of Cross-Complainant's rights and with the intent to vex, injure, or annoy Cross-Complainant, and were an intentional misrepresentation, deceit, or concealment of a material fact known to the Cross-Defendants with the intention to deprive Cross-Complainant of Property, legal rights or to otherwise cause injury, such as to constitute despicable conduct, oppression, fraud, or malice under California Civil Code Section 3294, entitling Cross-Complainant to punitive damages in an amount appropriate to punish or set an example of Cross-Defendants.

WHEREFORE, Cross-Complainant prays for judgment against Cross-Defendants, and each of them, as follows:

1. That the court determines the rights, duties and obligations of the parties to this action;
2. That in the event that it is determined that there is any sum due Cross-Complainant, that the court further declare that such liability be that of the Cross-Defendants, and that said Cross-Defendants are obligated to indemnify Cross-Complainant;
3. For total indemnity by Cross-Defendants in a sum to be established at the time of trial;
4. For the sum of $75,320.00 together with interest at the rate of ten per cent (10%) per annum;
5. Compensatory damages according to proof;
6. For punitive and exemplary damages;
7. For reasonable attorney's fees pursuant to law and/or contract;
8. For costs of suit incurred;
9. Or such other and further relief as the court deems just and proper.

Dated: _____,   The Mann Law Firm

                                                      By: _____

                                                                  Robert S. Mann

Attorneys for Defendants and Cross-Complainants, Defective Construction, Inc.

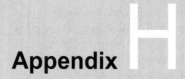
CALIFORNIA CODES

Civil Code Section 895

895. (a) "Structure" means any residential dwelling, other building, or improvement located upon a lot or within a common area.

(b) "Designed moisture barrier" means an installed moisture barrier specified in the plans and specifications, contract documents, or manufacturer's recommendations.

(c) "Actual moisture barrier" means any component or material, actually installed, that serves to any degree as a barrier against moisture, whether or not intended as such.

(d) "Unintended water" means water that passes beyond, around, or through a component or the material that is designed to prevent that passage.

(e) "Close of escrow" means the date of the close of escrow between the builder and the original homeowner. With respect to claims by an association, as defined in subdivision (a) of Section1351, "close of escrow" means the date of substantial completion, as defined in Section 337.15 of the Code of Civil Procedure, or the date the builder relinquishes control over the association's ability to decide whether to initiate a claim under this title, whichever is later.

(f) "Claimant" or "homeowner" includes the individual owners of single-family homes, individual unit owners of attached dwellings and, in the case of a common interest Development, any association as defined in subdivision (a) of Section 1351.

Civil Code Sections 896-897

896. In any action seeking recovery of damages arising out of, or related to deficiencies in, the residential construction, design, specifications, surveying, planning, supervision, testing, or observation of construction, a builder, and to the extent set forth in Chapter 4 (commencing with Section 910), a general contractor, subcontractor, material supplier, individual product manufacturer, or design professional, shall, except as specifically set forth in this title, be liable for, and the claimant's claims or causes of action shall be limited to violation of, the following standards, except as specifically set forth in this title. This title applies to original construction intended to be sold as an individual dwelling unit.

As to condominium conversions, this title does not apply to or does not supersede any other statutory or common law.

(a) With respect to water issues:

    1. A door shall not allow unintended water to pass beyond, around, or through the door or its designed or actual moisture barriers, if any.

2. Windows, patio doors, deck doors, and their systems shall not allow water to pass beyond, around, or through the window, patio door, or deck door or its designed or actual moisture barriers, including, without limitation, internal barriers within the systems themselves. For purposes of this paragraph, "systems" include, without limitation, windows, window assemblies, framing, substrate, flashings, and trim, if any.

3. Windows, patio doors, deck doors, and their systems shall not allow excessive condensation to enter the structure and cause damage to another component. For purposes of this paragraph, "systems" include, without limitation, windows, window assemblies, framing, substrate, flashings, and trim, if any.

4. Roofs, roofing systems, chimney caps, and ventilation components shall not allow water to enter the structure or to pass beyond, around, or through the designed or actual moisture barriers, including, without limitation, internal barriers located within the systems themselves. For purposes of this paragraph, "systems" include, without limitation, framing, substrate, and sheathing, if any.

5. Decks, deck systems, balconies, balcony systems, exterior stairs, and stair systems shall not allow water to pass into the adjacent structure. For purposes of this paragraph, "systems" include, without limitation, framing, substrate, flashing, and sheathing, if any.

6. Decks, deck systems, balconies, balcony systems, exterior stairs, and stair systems shall not allow unintended water to pass within the systems themselves and cause damage to the systems. For purposes of this paragraph, "systems" include, without limitation, framing, substrate, flashing, and sheathing, if any.

7. Foundation systems and slabs shall not allow water or vapor to enter into the structure so as to cause damage to another building component.

8. Foundation systems and slabs shall not allow water or vapor to enter into the structure so as to limit the installation of the type of flooring materials typically used for the particular application.

9. Hardscape, including paths and patios, irrigation systems, landscaping systems, and drainage systems, that are installed as part of the original construction, shall not be installed in such a way as to cause water or soil erosion to enter into or come in contact with the structure so as to cause damage to another building component.

10. Stucco, exterior siding, exterior walls, including, without limitation, exterior framing, and other exterior wall finishes and fixtures and the systems of those components and fixtures, including, but not limited to, pot shelves, horizontal surfaces, columns, and plant-ons, shall be installed in such a way so as not to allow unintended water to pass into the structure or to pass beyond, around, or through the designed or actual moisture barriers of the system, including any internal barriers located within the system itself. For purposes of this paragraph, "systems" include, without limitation, framing, substrate, flashings, trim, wall assemblies, and internal wall cavities, if any.

11. Stucco, exterior siding, and exterior walls shall not allow excessive condensation to enter the structure and cause damage to another component. For purposes of this paragraph, "systems" include, without limitation, framing, substrate, flashings, trim, wall assemblies, and internal wall cavities, if any.

12. Retaining and site walls and their associated drainage systems shall not allow unintended water to pass beyond, around, or through its designed or actual moisture barriers including, without limitation, any internal barriers, so as to cause damage. This standard does not apply to those portions of any wall or drainage system that are designed to have water flow beyond, around, or through them.

13. Retaining walls and site walls, and their associated drainage systems, shall only allow water to flow beyond, around, or through the areas designated by design.
14. The lines and components of the plumbing system, sewer system, and utility systems shall not leak.
15. Plumbing lines, sewer lines, and utility lines shall not corrode so as to impede the useful life of the systems.
16. Sewer systems shall be installed in such a way as to allow the designated amount of sewage to flow through the system.
17. Shower and bath enclosures shall not leak water into the interior of walls, flooring systems, or the interior of other components.
18. Ceramic tile and tile countertops shall not allow water into the interior of walls, flooring systems, or other components so as to cause damage.

(b) With respect to structural issues:
1. Foundations, load bearing components, and slabs, shall not contain significant cracks or significant vertical displacement.
2. Foundations, load bearing components, and slabs shall not cause the structure, in whole or in part, to be structurally unsafe.
3. Foundations, load bearing components, and slabs, and underlying soils shall be constructed so as to materially comply with the design criteria set by applicable government building codes, regulations, and ordinances for chemical deterioration or corrosion resistance in effect at the time of original construction.
4. A structure shall be constructed so as to materially comply with the design criteria for earthquake and wind load resistance, asset forth in the applicable government building codes, regulations, and ordinances in effect at the time of original construction.

(c) With respect to soil issues:
1. Soils and engineered retaining walls shall not cause, in whole or in part, damage to the structure built upon the soil or engineered retaining wall.
2. Soils and engineered retaining walls shall not cause, in whole or in part, the structure to be structurally unsafe.
3. Soils shall not cause, in whole or in part, the land upon which no structure is built to become unusable for the purpose represented at the time of original sale by the builder or for the purpose for which that land is commonly used.

(d) With respect to fire protection issues:
1. A structure shall be constructed so as to materially comply with the design criteria of the applicable government building codes, regulations, and ordinances for fire protection of the occupants ineffect at the time of the original construction.
2. Fireplaces, chimneys, chimney structures, and chimney termination caps shall be constructed and installed in such a way so as not to cause an unreasonable risk of fire outside the fireplace enclosure or chimney.
3. Electrical and mechanical systems shall be constructed and installed in such a way so as not to cause an unreasonable risk of fire.

(e) With respect to plumbing and sewer issues:

Plumbing and sewer systems shall be installed to operate properly and shall not materially impair the use of the structure by its inhabitants. However, no action may be brought for a violation of this subdivision more than four years after close of escrow.

(f) With respect to electrical system issues:

Electrical systems shall operate properly and shall not materially impair the use of the structure by its inhabitants. However, no action shall be brought pursuant to this subdivision more than four years from close of escrow.

(g) With respect to issues regarding other areas of construction:

1. Exterior pathways, driveways, hardscape, sidewalls, sidewalks, and patios installed by the original builder shall not contain cracks that display significant vertical displacement or that are excessive. However, no action shall be brought upon a violation of this paragraph more than four years from close of escrow.

2. Stucco, exterior siding, and other exterior wall finishes and fixtures, including, but not limited to, pot shelves, horizontal surfaces, columns, and plant-ons, shall not contain significant cracks or separations.

3a. To the extent not otherwise covered by these standards, manufactured products, including, but not limited to, windows, doors, roofs, plumbing products and fixtures, fireplaces, electrical fixtures, HVAC units, countertops, cabinets, paint, and appliances shall be installed so as not to interfere with the products' useful life, if any.

3b. For purposes of this paragraph, "useful life" means are presentation of how long a product is warranted or represented, through its limited warranty or any written representations, to last by its manufacturer, including recommended or required maintenance. If there is no representation by a manufacturer, a builder shall install manufactured products so as not to interfere with the product's utility.

3c. For purposes of this paragraph, "manufactured product" means a product that is completely manufactured offsite.

3d. If no useful life representation is made, or if there presentation is less than one year, the period shall be no less than one year. If a manufactured product is damaged as a result of a violation of these standards, damage to the product is a recoverable element of damages. This subparagraph does not limit recovery if there has been damage to another building component caused by a manufactured product during the manufactured product's useful life.

3e. This title does not apply in any action seeking recovery solely for a defect in a manufactured product located within or adjacent to a structure.

4. Heating, if any, shall be installed so as to be capable of maintaining a room temperature of 70 degrees Fahrenheit at a point three feet above the floor in any living space.

5. Living space air-conditioning, if any, shall be provided in a manner consistent with the size and efficiency design criteria specified in Title 24 of the California Code of Regulations or its successor.

6. Attached structures shall be constructed to comply with inter unit noise transmission standards set by the applicable government building codes, ordinances, or regulations in effect at the time of the original construction. If there is no applicable code, ordinance, or regulation, this paragraph does not apply. However, no action shall be brought pursuant to this paragraph more than one year from the original occupancy of the adjacent unit.

7. Irrigation systems and drainage shall operate properly so as not to damage landscaping or other external improvements. However, no action shall be brought pursuant to this paragraph more than one year from close of escrow.

8. Untreated wood posts shall not be installed in contact with soil so as to cause unreasonable decay to the wood based upon the finish grade at the time of original construction. However, no action shall be brought pursuant to this paragraph more than two years from close of escrow.

9. Untreated steel fences and adjacent components shall be installed so as to prevent unreasonable corrosion. However, no action shall be brought pursuant to this paragraph more than four years from close of escrow.

10. Paint and stains shall be applied in such a manner so as not to cause deterioration of the building surfaces for the length of time specified by the paint or stain manufacturers' representations, if any. However, no action shall be brought pursuant to this paragraph more than five years from close of escrow.

11. Roofing materials shall be installed so as to avoid materials falling from the roof.

12. The landscaping systems shall be installed in such a manner so as to survive for not less than one year. However, no action shall be brought pursuant to this paragraph more than two years from close of escrow.

13. Ceramic tile and tile backing shall be installed in such a manner that the tile does not detach.

14. Dryer ducts shall be installed and terminated pursuant to manufacturer installation requirements. However, no action shall be brought pursuant to this paragraph more than two years from close of escrow.

15. Structures shall be constructed in such a manner so as not to impair the occupants' safety because they contain public health hazards as determined by a duly authorized public health official, health agency, or governmental entity having jurisdiction.

This paragraph does not limit recovery for any damages caused by a violation of any other paragraph of this section on the grounds that the damages do not constitute a health hazard.

897. The standards set forth in this chapter are intended to address every function or component of a structure. To the extent that a function or component of a structure is not addressed by these standards, it shall be actionable if it causes damage.

Civil Code Sections 900-907

900. As to fit and finish items, a builder shall provide a home buyer with a minimum one-year express written limited warranty covering the fit and finish of the following building components. Except as otherwise provided by the standards specified in Chapter 2 (commencing with Section 896), this warranty shall cover the fit and finish of cabinets, mirrors, flooring, interior and exterior walls, countertops, paint finishes, and trim, but shall not apply to damage to those components caused by defects in other components governed by the other provisions of this title. Any fit and finish matters covered by this warranty are not subject to the provisions of this title. If a builder fails to provide the express warranty required by this section, the warranty for these items shall be for a period of one year.

901. A builder may, but is not required to, offer greater protection or protection for longer time periods in its express contract with the homeowner than that set forth in Chapter 2 (commencing with Section 896). A builder may not limit the application of Chapter 2 (commencing with Section 896) or lower its protection through the express contract with the homeowner. This type of express contract constitutes an "enhanced protection agreement."

902. If a builder offers an enhanced protection agreement, the builder may choose to be subject to its own express contractual provisions in place of the provisions set forth in Chapter 2(commencing with Section 896). If an enhanced protection agreement is in place, Chapter 2 (commencing with Section 896) no longer applies other than to set forth minimum provisions by which to judge the enforceability of the particular provisions of the enhanced protection agreement.

903. If a builder offers an enhanced protection agreement in place of the provisions set forth in Chapter 2 (commencing with Section 896), the election to do so shall be made in writing with the homeowner no later than the close of escrow. The builder shall provide the homeowner with a complete copy of Chapter 2 (commencing with Section 896) and advise the homeowner that the builder has elected not to be subject to its provisions. If any provision of an enhanced protection agreement is later found to be unenforceable as not meeting the minimum standards of Chapter 2 (commencing with Section 896), a builder may use this chapter in lieu of those provisions found to be unenforceable.

904. If a builder has elected to use an enhanced protection agreement, and a homeowner disputes that the particular provision or time periods of the enhanced protection agreement are not greater than, or equal to, the provisions of Chapter 2 (commencing with Section 896) as they apply to the particular deficiency alleged by the homeowner, the homeowner may seek to enforce the application of the standards set forth in this chapter as to those claimed deficiencies. If a homeowner seeks to enforce a particular standard in lieu of a provision of the enhanced protection agreement, the homeowner shall give the builder written notice of that intent at the time the homeowner files a notice of claim pursuant to Chapter 4 (commencing with Section 910).

905. If a homeowner seeks to enforce Chapter 2 (commencing with Section 896), in lieu of the enhanced protection agreement in a subsequent litigation or other legal action, the builder shall have the right to have the matter bifurcated, and to have an immediately binding determination of his or her responsive pleading within 60 days after the filing of that pleading, but in no event after the commencement of discovery, as to the application of either Chapter 2 (commencing with Section 896) or the enhanced protection agreement as to the deficiencies claimed by the homeowner. If the builder fails to seek that determination in the timeframe specified, the builder waives the right to do so and the standards set forth in this title shall apply. As to any non original homeowner, that homeowner shall be deemed in privity for purposes of an enhanced protection agreement only to the extent that the builder has recorded the enhanced protection agreement on title or provided actual notice to the non original homeowner of the enhanced protection agreement. If the enhanced protection agreement is not recorded

on title or no actual notice has been provided, the standards set forth in this title apply to any non original homeowners' claims.

906. A builder's election to use an enhanced protection agreement addresses only the issues set forth in Chapter 2 (commencing with Section 896) and does not constitute an election to use or not use the provisions of Chapter 4 (commencing with Section 910). The decision to use or not use Chapter 4 (commencing with Section 910) is governed by the provisions of that chapter.

907. A homeowner is obligated to follow all reasonable maintenance obligations and schedules communicated in writing to the homeowner by the builder and product manufacturers, as well as commonly accepted maintenance practices. A failure by a homeowner to follow these obligations, schedules, and practices may subject the homeowner to the affirmative defenses contained in Section 944.

Civil Code Sections 910-938

910. Prior to filing an action against any party alleged to have contributed to a violation of the standards set forth in Chapter 2 (commencing with Section 896), the claimant shall initiate the following prelitigation procedures:

(a) The claimant or his or her legal representative shall provide written notice via certified mail, overnight mail, or personal delivery to the builder, in the manner prescribed in this section, of the claimant's claim that the construction of his or her residence violates any of the standards set forth in Chapter 2 (commencing with Section 896). That notice shall provide the claimant's name, address, and preferred method of contact, and shall state that the claimant alleges a violation pursuant to this part against the builder, and shall describe the claim in reasonable detail sufficient to determine the nature and location, to the extent known, of the claimed violation. In the case of a group of homeowners or an association, the notice may identify the claimants solely by addressor other description sufficient to apprise the builder of the locations of the subject residences. That document shall have the same force and effect as a notice of commencement of a legal proceeding.

(b) The notice requirements of this section do not preclude a homeowner from seeking redress through any applicable normal customer service procedure as set forth in any contractual, warranty, or other builder-generated document; and, if a homeowner seeks to do so, that request shall not satisfy the notice requirements of this section.

911. (a) For purposes of this title, except as provided in subdivision

(b) "builder" means any entity or individual, including, but not limited to a builder, developer, general contractor, contractor, or original seller, who, at the time of sale, was also inthe business of selling residential units to the public for the property that is the subject of the homeowner's claim or was in the business of building, developing, or constructing residential units for public purchase for the Property that is the subject of the homeowner's claim.

(c) For the purposes of this title, "builder" does not include any entity or individual whose involvement with a residential unit that is the subject of the homeowner's claim is limited to his or her capacity as general contractor or contractor and who is not a

324 | Defect-Free Buildings: A Construction Manual for Quality Control and Conflict Resolution

partner, member of, subsidiary of, or otherwise similarly affiliated with the builder. For purposes of this title, these non affiliated general contractors and nonaffiliated contractors shall be treated the same as subcontractors, material suppliers, individual product manufacturers, and design professionals.

912. A builder shall do all of the following:

(a) Within 30 days of a written request by a homeowner or his or her legal representative, the builder shall provide copies of all relevant plans, specifications, mass or rough grading plans, final soils reports, Department of Real Estate public reports, and available engineering calculations, that pertain to a homeowner's residence specifically or as part of a larger Development tract. The request shall be honored if it states that it is made relative to structural, fire safety, or soils provisions of this title. However, a builder is not obligated to provide a copying service, and reasonable copying costs shall be borne by the requesting party. A builder may require that the documents be copied onsite by the requesting party, except that the homeowner may, at his or her option, use his or her own copying service, which may include an offsite copy facility that is bonded and insured. If a builder can show that the builder maintained the documents, but that they later became unavailable due to loss or destruction that was not the fault of the builder, the builder may be excused from the requirements of this subdivision, in which case the builder shall act with reasonable diligence to assist the homeowner in obtaining those documents from any applicable government authority or from the source that generated the document. However, in that case, the time limits specified by this section do not apply.

(b) At the expense of the homeowner, who may opt to use an off site copy facility that is bonded and insured, the builder shall provide to the homeowner or his or her legal representative copies of all maintenance and preventative maintenance recommendations that pertain to his or her residence within 30 days of service of a written request for those documents. Those documents shall also be provided to the homeowner in conjunction with the initial sale of the residence.

(c) At the expense of the homeowner, who may opt to use an off site copy facility that is bonded and insured, a builder shall provide to the homeowner or his or her legal representative copies of all manufactured products maintenance, preventive maintenance, and limited warranty information within 30 days of a written request for those documents. These documents shall also be provided to the homeowner in conjunction with the initial sale of the residence.

(d) At the expense of the homeowner, who may opt to use an off site copy facility that is bonded and insured, a builder shall provide to the homeowner or his or her legal representative copies of all of the builder's limited contractual warranties in accordance with this part in effect at the time of the original sale of the residence within 30 days of a written request for those documents. Those documents shall also be provided to the homeowner in conjunction with the initial sale of the residence.

(e) A builder shall maintain the name and address of an agent for notice pursuant to this chapter with the Secretary of State or, alternatively, elect to use a third party for that notice if the builder has notified the homeowner in writing of the third party's name and address, to whom claims and requests for information under this section may be mailed. The name and address of the agent for notice or third party shall be included with the original sales documentation and shall be initialed and acknowledged by the purchaser and the builder's sales representative. This subdivision applies to instances in which a builder contracts with a third party to accept claims and act on the builder's behalf. A builder shall give actual notice to the homeowner that the builder has made such an election, and shall include the name and address of the third party.

(f) A builder shall record on title a notice of the existence of these procedures and a notice that these procedures impact the legal rights of the homeowner. This information shall also be included with the original sales documentation and shall be initialed and acknowledged by the purchaser and the builder's sales representative.

(g) A builder shall provide, with the original sales documentation, a written copy of this title, which shall be initialed and acknowledged by the purchaser and the builder's sales representative.

(h) As to any documents provided in conjunction with the original sale, the builder shall instruct the original purchaser to provide those documents to any subsequent purchaser.

(i) Any builder who fails to comply with any of these requirements within the time specified is not entitled to the protection of this chapter, and the homeowner is released from the requirements of this chapter and may proceed with the filing of an action, in which case the remaining chapters of this part shall continue to apply to the action.

913. A builder or his or her representative shall acknowledge, in writing, receipt of the notice of the claim within 14 days after receipt of the notice of the claim. If the notice of the claim is served by the claimant's legal representative, or if the builder receives a written representation letter from a homeowner's attorney, the builder shall include the attorney in all subsequent substantive communications, including, without limitation, all written communications occurring pursuant to this chapter, and all substantive and procedural communications, including all written communications, following the commencement of any subsequent complaint or other legal action, except that if the builder has retained or involved legal counsel to assist the builder in this process, all communications by the builder's counsel shall only be with the claimant's legal representative, if any.

914. (a) This chapter establishes a nonadversarial procedure, including the remedies available under this chapter which, if the procedure does not resolve the dispute between the parties, may result in a subsequent action to enforce the other chapters of this title. A builder may attempt to commence nonadversarial contractual provisions other than the nonadversarial procedures and remedies set forth in this chapter, but may not, in addition to its own nonadversarial contractual provisions, require adherence to the nonadversarial procedures and remedies set forth in this chapter, regardless of whether the builder's own alternative nonadversarial contractual provisions are successful in resolving the dispute or ultimately deemed enforceable. At the time the sales agreement is executed, the builder shall notify the homeowner whether the builder intends to engage in the nonadversarial procedure of this section or attempt to enforce alternative nonadversarial contractual provisions. If the builder elects to use alternative nonadversarial contractual provisions in lieu of this chapter, the election is binding, regardless of whether the builder's alternative nonadversarial contractual provisions are successful in resolving the ultimate dispute or are ultimately deemed enforceable.

(b) Nothing in this title is intended to affect existing statutory or decisional law pertaining to the applicability, viability, or enforceability of alternative dispute resolution methods, alternative remedies, or contractual arbitration, judicial reference, or similar procedures requiring a binding resolution to enforce the other chapters of this title or any other disputes between homeowners and builders. Nothing in this title is intended to affect the applicability, viability, or enforceability, if any, of contractual arbitration or judicial reference after a nonadversarial procedure or provision has been completed.

915. If a builder fails to acknowledge receipt of the notice of a claim within the time specified, elects not to go through the process set forth in this chapter, or fails to request an inspection within the time specified, or at the conclusion or cessation of an alternative nonadversarial proceeding, this chapter does not apply and the homeowner is released from the requirements of this chapter and may proceed with the filing of an action. However, the standards set forth in the other chapters of this title shall continue to apply to the action.

916. (a) If a builder elects to inspect the claimed unmet standards, the builder shall complete the initial inspection and testing within 14 days after acknowledgment of receipt of the notice of the claim, at a mutually convenient date and time. If the homeowner has retained legal representation, the inspection shall be scheduled with the legal representative's office at a mutually convenient date and time, unless the legal representative is unavailable during the relevant time periods. All costs of builder inspection and testing, including any damage caused by the builder inspection, shall be borne by the builder. The builder shall also provide written proof that the builder has liability insurance to cover any damages or injuries occurring during inspection and testing. The builder shall restore the Property to its pre testing condition within 48 hours of the testing. The builder shall, upon request, allow the inspections to be observed and electronically recorded, videotaped, or photographed by the claimant or his or her legal representative.

(b) Nothing that occurs during a builder's or claimant's inspection or testing may be used or introduced as evidence to support a spoliation defense by any potential party in any subsequent litigation.

(c) If a builder deems a second inspection or testing reasonably necessary, and specifies the reasons therefore in writing within three days following the initial inspection, the builder may conduct a second inspection or testing. A second inspection or testing shall be completed within 40 days of the initial inspection or testing. All requirements concerning the initial inspection or testing shall also apply to the second inspection or testing.

(d) If the builder fails to inspect or test the Property within the time specified, the claimant is released from the requirements of this section and may proceed with the filing of an action. However, the standards set forth in the other chapters of this title shall continue to apply to the action.

(e) If a builder intends to hold a subcontractor, design professional, individual product manufacturer, or material supplier, including an insurance carrier, warranty company, or service company, responsible for its contribution to the unmet standard, the builder shall provide notice to that person or entity sufficiently in advance to allow them to attend the initial, or if requested, second inspection of any alleged unmet standard and to participate in the repair process. The claimant and his or her legal representative, if any, shall be advised in a reasonable time prior to the inspection as to the identity of all persons or entities invited to attend. This subdivision does not apply to the builder's insurance company. Except with respect to any claims involving a repair actually conducted under this chapter, nothing in this subdivision shall be construed to relieve a subcontractor, design professional, individual product manufacturer, or material supplier of any liability under an action brought by a claimant.

917. Within 30 days of the initial or, if requested, second inspection or testing, the builder may offer in writing to repair the violation. The offer to repair shall also compensate the homeowner for all applicable damages recoverable under Section 944, within the timeframe for the repair set forth in this chapter. Any such offer shall be accompanied by a detailed, specific, step-by-step statement identifying the particular violation that is being repaired, explaining the nature, scope, and location of the repair, and setting a reasonable completion date for the repair. The offer shall also include the names, addresses, telephone numbers, and license numbers of the contractors whom the builder intends to have to perform the repair. Those contractors shall be fully insured for, and shall be responsible for, all damages or injuries that they may cause to occur during the repair, and evidence of that insurance shall be provided to the homeowner upon request. Upon written request by the homeowner or his or her legal representative, and within the timeframes set forth in this chapter, the builder shall also provide any available technical documentation, including, without limitation, plans and specifications, pertaining to the claimed violation within the particular home or Development tract. The offer shall also advise the homeowner in writing of his or her right to request up to three additional contractors from which to select to do the repair pursuant to this chapter.

918. Upon receipt of the offer to repair, the homeowner shall have 30 days to authorize the builder to proceed with the repair. The homeowner may alternatively request, at the homeowner's sole option and discretion, that the builder provide the names, addresses, telephone numbers, and license numbers for up to three alternative contractors who are not owned or financially controlled by the builder and who regularly conduct business in the county where the structure is located. If the homeowner so elects, the builder is entitled to an additional noninvasive inspection, to occur at a mutually convenient date and time within 20 days of the election, so as to permit the other proposed contractors to review the proposed site of the repair. Within 35 days after the request of the homeowner for alternative contractors, the builder shall present the homeowner with a choice of contractors. Within 20 days after that presentation, the homeowner shall authorize the builder or one of the alternative contractors to perform the repair.

919. The offer to repair shall also be accompanied by an offer to mediate the dispute if the homeowner so chooses. The mediation shall be limited to a four-hour mediation, except as otherwise mutually agreed before a nonaffiliated mediator selected and paid for by the builder. At the homeowner's sole option, the homeowner may agree to split the cost of the mediator, and if he or she does so, the mediator shall be selected jointly. The mediator shall have sufficient availability such that the mediation occurs within 15 days after the request to mediate is received and occurs at a mutually convenient location within the county where the action is pending. If a builder has made an offer to repair a violation, and the mediation has failed to resolve the dispute, the homeowner shall allow the repair to be performed either by the builder, its contractor, or the selected contractor.

920. If the builder fails to make an offer to repair or otherwise strictly comply with this chapter within the times specified, the claimant is released from the requirements of this chapter and may proceed with the filing of an action. If the contractor performing the repair does not complete the repair in the time or manner specified, the claimant may file an action. If this occurs, the standards set forth in the other chapters of this part shall continue to apply to the action.

921. (a) In the event that a resolution under this chapter involves a repair by the builder, the builder shall make an appointment with the claimant, make all appropriate arrangements to effectuate a repair of the claimed unmet standards, and compensate the homeowner for all damages resulting there from free of charge to the claimant. The repair shall be scheduled through the claimant's legal representative, if any, unless he or she is unavailable during the relevant time periods. The repair shall be commenced on a mutually convenient date within 14 days of acceptance or, if an alternative contractor is selected by the homeowner, within 14 days of the selection, or, if a mediation occurs, within seven days of the mediation, or within five days after a permit is obtained if one is required. The builder shall act with reasonable diligence in obtaining any such permit.

(b) The builder shall ensure that work done on the repairs is done with the utmost diligence, and that the repairs are completed as soon as reasonably possible, subject to the nature of the repair or some unforeseen event not caused by the builder or the contractor performing the repair. Every effort shall be made to complete the repair within 120 days.

922. The builder shall, upon request, allow the repair to be observed and electronically recorded, videotaped, or photographed by the claimant or his or her legal representative. Nothing that occurs during the repair process may be used or introduced as evidence to support a spoliation defense by any potential party in any subsequent litigation.

923. The builder shall provide the homeowner or his or her legal representative, upon request, with copies of all correspondence, photographs, and other materials pertaining or relating in any manner to the repairs.

924. If the builder elects to repair some, but not all of, the claimed unmet standards, the builder shall, at the same time it makes its offer, set forth with particularity in writing the reasons, and the support for those reasons, for not repairing all claimed unmet standards.

925. If the builder fails to complete the repair within the time specified in the repair plan, the claimant is released from the requirements of this chapter and may proceed with the filing of an action. If this occurs, the standards set forth in the other chapters of this title shall continue to apply to the action.

926. The builder may not obtain a release or waiver of any kind in exchange for the repair work mandated by this chapter. At the conclusion of the repair, the claimant may proceed with filing an action for violation of the applicable standard or for a claim of inadequate repair, or both, including all applicable damages available under Section 944.

927. If the applicable statute of limitations has otherwise run during this process, the time period for filing a complaint or other legal remedies for violation of any provision of this title, or for a claim of inadequate repair, is extended from the time of the original claim by the claimant to 100 days after the repair is completed, whether or not the particular violation is the one being repaired. If the builder fails to acknowledge the

claim within the time specified, elects not to go through this statutory process, or fails to request an inspection within the time specified, the time period for filing a complaint or other legal remedies for violation of any provision of this title is extended from the time of the original claim by the claimant to 45 days after the time for responding to the notice of claim has expired. If the builder elects to attempt to enforce its own nonadversarial procedure in lieu of the procedure set forth in this chapter, the time period for filing a complaint or other legal remedies for violation of any provision of this part is extended from the time of the original claim by the claimant to 100 days after either the completion of the builder's alternative nonadversarial procedure, or 100 days after the builder's alternative nonadversarial procedure is deemed unenforceable, whichever is later.

928. If the builder has invoked this chapter and completed a repair, prior to filing an action, if there has been no previous mediation between the parties, the homeowner or his or her legal representative shall request mediation in writing. The mediation shall be limited to four hours, except as otherwise mutually agreed before a nonaffiliated mediator selected and paid for by the builder. At the homeowner's sole option, the homeowner may agree to split the cost of the mediator and if he or she does so, the mediator shall be selected jointly. The mediator shall have sufficient availability such that the mediation will occur within 15 days after the request for mediation is received and shall occur at a mutually convenient location within the county where the action is pending. In the event that a mediation is used at this point, any applicable statutes of limitations shall be tolled from the date of the request to mediate until the next court day after the mediation is completed, or the 100 day period, whichever is later.

929. (a) Nothing in this chapter prohibits the builder from making only a cash offer and no repair. In this situation, the homeowner is free to accept the offer, or he or she may reject the offer and proceed with the filing of an action. If the latter occurs, the standards of the other chapters of this title shall continue to apply to the action.

(b) The builder may obtain a reasonable release in exchange for the cash payment. The builder may negotiate the terms and conditions of any reasonable release in terms of scope and consideration in conjunction with a cash payment under this chapter.

930. (a) The time periods and all other requirements in this chapter are to be strictly construed, and, unless extended by the mutual agreement of the parties in accordance with this chapter, shall govern the rights and obligations under this title. If a builder fails to act in accordance with this section within the time frames mandated, unless extended by the mutual agreement of the parties as evidenced by a post claim written confirmation by the affected homeowner demonstrating that he or she has knowingly and voluntarily extended the statutory timeframe, the claimant may proceed with filing an action. If this occurs, the standards of the other chapters of this title shall continue to apply to the action.

(b) If the claimant does not conform with the requirements of this chapter, the builder may bring a motion to stay any subsequent court action or other proceeding until the requirements of this chapter have been satisfied. The court, in its discretion, may award the prevailing party on such a motion, his or her attorney's fees and costs in bringing or opposing the motion.

931. If a claim combines causes of action or damages not covered by this part, including, without limitation, personal injuries, class actions, other statutory remedies, or fraud-based claims, the claimed unmet standards shall be administered according to this part, although evidence of the Property in its unrepaired condition may be introduced to support the respective elements of any such cause of action. As to any fraud-based claim, if the fact that the Property has been repaired under this chapter is deemed admissible, the trier of fact shall be informed that the repair was not voluntarily accepted by the homeowner. As to any class action claims that address solely the incorporation of a defective component into a residence, the named and unnamed class members need not comply with this chapter.

932. Subsequently discovered claims of unmet standards shall be administered separately under this chapter, unless otherwise agreed to by the parties. However, in the case of a detached single family residence, in the same home, if the subsequently discovered claim is for a violation of the same standard as that which has already been initiated by the same claimant and the subject of a currently pending action, the claimant need not reinitiate the process as to the same standard. In the case of an attached project, if the subsequently discovered claim is for a violation of the same standard for a connected component system in the same building as has already been initiated by the same claimant, and the subject of a currently pending action, the claimant need not reinitiate this process as to that standard.

933. If any enforcement of these standards is commenced, the fact that a repair effort was made may be introduced to the trier of fact. However, the claimant may use the condition of the Property prior to the repair as the basis for contending that the repair work was inappropriate, inadequate, or incomplete, or that the violation still exists. The claimant need not show that the repair work resulted in further damage nor that damage has continued to occur as a result of the violation.

934. Evidence of both parties' conduct during this process may be introduced during a subsequent enforcement action, if any, with the exception of any mediation. Any repair efforts undertaken by the builder, shall not be considered settlement communications or offers of settlement and are not inadmissible in evidence on such a basis.

935. To the extent that provisions of this chapter are enforced and those provisions are substantially similar to provisions in Section1375 of the Civil Code, but an action is subsequently commenced under Section 1375 of the Civil Code, the parties are excused from performing the substantially similar requirements under Section 1375 of the Civil Code.

936. Each and every provision of the other chapters of this title apply to general contractors, subcontractors, material suppliers, individual product manufacturers, and design professionals to the extent that the general contractors, subcontractors, material suppliers, individual product manufacturers, and design professionals caused, in whole or in part, a violation of a particular standard as the result of a negligent act or omission or a breach of contract. In addition to the affirmative defenses set forth in Section 945.5, a general contractor, subcontractor, material supplier, design professional, individual product manufacturer, or other entity may also offer common law and contractual defenses as applicable to any claimed violation of a standard. All actions by a claimant or builder to enforce an express contract, or any provision there of, against a general contractor,

subcontractor, material supplier, individual product manufacturer, or design professional is preserved. Nothing in this title modifies the law pertaining to joint and several liabilities for builders, general contractors, subcontractors, material suppliers, individual product manufacturer, and design professionals that contribute to any specific violation of this title. However, the negligence standard in this section does not apply to any general contractor, subcontractor, material supplier, individual product manufacturer, or design professional with respect to claims for which strict liability would apply.

937. Nothing in this title shall be interpreted to eliminate or abrogate the requirement to comply with Section 411.35 of the Code of Civil Procedure or to affect the liability of design professionals, including architects and architectural firms, for claims and damages not covered by this title.

938. This title applies only to new residential units where the purchase agreement with the buyer was signed by the seller on or after January 1, 2003.

Bibliography

Acret, James. 1985. *Construction Arbitration Handbook*. Colorado Springs, CO: Shepard's/McGraw-Hill.

—. 1986. *Construction Litigation Handbook*. Colorado Springs, CO: Shepard's/McGraw-Hill.

—. 1990. *Attorney's Guide to California Construction Contracts and Disputes*, 2nd Edition. Berkeley, CA: Continuing Education of the Bar, California.

—. 1990. *California Construction Law Manual*. New York: McGraw-Hill.

Ambrose, James, and Dimitry Vergun. 1999. *Design for Earthquakes*. New York: John Wiley & Sons.

Bliss, Steven, ed. 2003. *Troubleshooting Guide to Residential Construction: The Diagnosis and Prevention of Common Building Problems*. Williston, VT: The Journal of Light Construction.

Board of Health Promotion and Disease Prevention, Institute of Medicine of the National Academies Committee on Damp Indoor Spaces and Health. 2004. *Damp Indoor Spaces and Health*. Washington, D.C.: The National Academies Press.

Breyer, Donald F., Kenneth J. Fridley, and Kelly E. Cobeen. 1999. *Design of Wood Structures*, 4th Edition. New York: McGraw-Hill.

Bundschuh, Gregg. 1997. *Owner's Guide to Construction Risk Management and Insurance*. Atlanta, GA: J&H Marsh &McClennan.

Costello, Edward J., Jr. 1996. *Controlling Conflict*. Chicago: CCH Incorporated.

Cushman, Kenneth M., ed. 1981. *Construction Litigation*. New York: Practicing Law Institute.

Cushman, Robert F., and David A. Carpenter, eds. 1990. *Proving and Pricing Construction Claims*. New York: Wiley Law Publications, John Wiley & Sons, Inc.

Cushman, Robert F., and Stephen D. Butler, eds. 1994. *Construction Change Order Claims*. New York: Wiley Law Publications, John Wiley & Sons, Inc.

DeKorne, Clayton, ed. 2003. *JLC Field Guide to Residential Construction: A Manual of Best Practice*. Williston, VT: Hanley-Wood, LLC.

Feld, Jacob, and Kenneth L. Carper. 1997. *Construction Failure*, 2nd Edition. New York: John Wiley & Sons, Inc.

Hardie, Glenn M. 1995. *Building Construction, Principles, Practices and Materials*. Englewood Cliffs, NJ: Prentice Hall.

Jaffe, David S. 1996. *Contracts and Liability*, 4th Edition. Washington, D.C.: Home Builder Press, National Association of Home Builders.

Jones, Glower W. 1998. *Alternative Clauses to Standard Construction Contracts*, 2nd Edition. Gaithersburg, MD: Aspen Law & Business.

Kubal, Michael T. 1993. *Waterproofing the Building Envelope*. New York: McGraw-Hill, Inc.

Lieff, M., and J.R. Treschel, eds. 1982. *Moisture Migration in Buildings, ASTM Special Technical Publication 779*. Philadelphia: American Society for Testing and Materials,

Moore, Christopher W. 1986. *The Mediation Process: Practical Solutions for Resolving Conflict*. San Francisco: Jossey-Bass, Inc.

O'Leary, Arthur F. 1999. *A Guide to Successful Construction*, 3rd Revised Edition. Los Angeles: BNI Building New Publications.

Olin, Harold B., John L. Schmidt, and Walter H. Lewis. 1990. *Construction: Principles, Materials and Methods*. New York: Van Nostrand Reinhold.

Sherwood, Gerald E., and Robert C. Stroh, eds. 1992. *Wood Frame House Construction*. Washington, D.C.: Home Builder Press, National Association of Home Builders.

Simmons, H. Leslie. 2001. *Construction: Principles, Materials, and Methods*, 7th Edition. New York: John Wiley and Sons, Inc.

Stein, Steven G.M., ed. 1992. *Construction Law*. New York: Matthew Bender.

Sweet, Justin. 1997. *Sweet on Construction Law*. Chicago: ABA Publishing, American Bar Association.

Wahlfeldt, Bette Galman. 1988. *Wood Frame Housebuilding*. Blue Ridge Summit, PA: Tab Books, Inc.

Whitney, Christopher C., Robert J. MacPherson, and James, Duffy, eds. 2001. *Sticks & Bricks*. Chicago: ABA Publishing, American Bar Association

Yanev, Peter I. 1991. *Peace of Mind in Earthquake Country*. San Francisco: Chronicle Books.

Index